Springer Theses

Recognizing Outstanding Ph.D. Research

Aims and Scope

The series "Springer Theses" brings together a selection of the very best Ph.D. theses from around the world and across the physical sciences. Nominated and endorsed by two recognized specialists, each published volume has been selected for its scientific excellence and the high impact of its contents for the pertinent field of research. For greater accessibility to non-specialists, the published versions include an extended introduction, as well as a foreword by the student's supervisor explaining the special relevance of the work for the field. As a whole, the series will provide a valuable resource both for newcomers to the research fields described, and for other scientists seeking detailed background information on special questions. Finally, it provides an accredited documentation of the valuable contributions made by today's younger generation of scientists.

Theses may be nominated for publication in this series by heads of department at internationally leading universities or institutes and should fulfill all of the following criteria

- They must be written in good English.
- The topic should fall within the confines of Chemistry, Physics, Earth Sciences, Engineering and related interdisciplinary fields such as Materials, Nanoscience, Chemical Engineering, Complex Systems and Biophysics.
- The work reported in the thesis must represent a significant scientific advance.
- If the thesis includes previously published material, permission to reproduce this must be gained from the respective copyright holder (a maximum 30% of the thesis should be a verbatim reproduction from the author's previous publications).
- They must have been examined and passed during the 12 months prior to nomination.
- Each thesis should include a foreword by the supervisor outlining the significance of its content.
- The theses should have a clearly defined structure including an introduction accessible to new PhD students and scientists not expert in the relevant field.

Indexed by zbMATH.

More information about this series at https://link.springer.com/bookseries/8790

Marina Trad Nery

Laser Power Stabilization via Radiation Pressure

Doctoral Thesis accepted by
The Leibniz University Hanover, Hanover, Germany

Springer

Author
Marina Trad Nery
Leibniz Universität Hannover
and Max-Planck Institut für
Gravitationsphysik
Hannover, Niedersachsen, Germany

Supervisor
Prof. Benno Willke
Leibniz Universität Hannover
Hannover, Niedersachsen, Germany

Max-Planck Institut für Gravitationsphysik
Hannover, Germany

ISSN 2190-5053 ISSN 2190-5061 (electronic)
Springer Theses
ISBN 978-3-030-95867-1 ISBN 978-3-030-95868-8 (eBook)
https://doi.org/10.1007/978-3-030-95868-8

This Springer imprint is published by the registered company Springer Nature Switzerland AG
The registered company address is: Gewerbestrasse 11, 6330 Cham, Switzerland

Supervisor's Foreword

Highly stable lasers are essential in many areas of experimental physics, several of which additionally require high laser power. Some examples are laser cooling setups in quantum optics or light-shining-through-wall experiments in particle physics. The most demanding requirements on the stability of high-power lasers, however, are set by laser interferometric gravitational wave detectors. Only the extremely high stability of the high-power laser system allowed the first detection of gravitational waves from merging black holes, a ground-breaking observation that was awarded the Nobel Prize in Physics in 2017. Next-generation gravitational wave detectors are expected to see events to the edge of the visible universe and might shed some light on the earliest times of our universe that can't be analyzed by electromagnetic radiation. These new instruments will require laser systems that are ten times more stable than the currently used ones. To achieve such stabilities, new directions must be taken, especially with respect to laser power stabilization. The main problem in power noise reduction is the photon shot noise in the power noise sensing process. The impact of this noise source can be reduced by increasing the power level on the power noise sensor which in turn is limited by the damage threshold of photodiodes in traditional power noise sensing schemes. One way to circumvent this problem is to transfer the power fluctuations to a different observable. This general idea is around for more than 15 years but was to my knowledge never quantitatively analyzed until the start of the thesis work by Marina Trad Nery. In her thesis, Marina Trad Nery analyses several transfer schemes of power fluctuations to different laser observables. She finds the transfer to phase fluctuations via radiation pressure to be the most promising concept given current technology and non-linear material parameters. In this scheme, laser power modulation is transferred via radiation pressure force to the motion of a micro-oscillator mirror. This motion is analyzed with a Michelson interferometer as the sensor of a power stabilization feedback control loop. Marina Trad Nery performs a detailed quantum optical analysis of this scheme and finds that, under the assumption of realistic experimental parameters, the generation of a non-classical squeezed light beam with bright coherent excitation should be possible. Not satisfied by the theoretical analysis, she performed a proof-of-principle experiment that achieved a remarkable sensitivity. The results of this experiment agreed very well with the theoretical predictions and were limited in a wide frequency range by the fundamental thermal noise of the used micro-oscillator mirror. The thesis is written in precise scientific language. The calculations and the experiment are described with sufficient detail to allow non-experts to follow the line of thought and to understand the main results. Many appropriate references are cited for a deeper understanding of the results and of the arguments given. The measurements and simulations are discussed in great detail to clarify their relevance and implications. The outstanding quality of Marina Trad Nery's work was recognized by the excellent mark "summa cum laude" and honored by the Prize of the Leibniz Universitätsgesellschaft e. V.. Furthermore, the Gravitational Wave International Committee (GWIC) recognized the importance of this work to the field of gravitational waves and gravitational wave detection by the 2020 GWIC-Braccini Thesis prize. It was a pleasure to accompany Marina Trad Nery's theoretical and experimental work and I very much enjoyed reading her thesis presented in this book.

Hannover, Germany
February 2022

Prof. Benno Willke

Preface

This work reports a new active power stabilization scheme which can be implemented in high-precision experiments, such as gravitational wave detectors. The novel aspect of the scheme is sensing laser power fluctuations via the radiation pressure driven motion they induce on a movable mirror. The mirror position and its fluctuations are determined by means of a weak auxiliary beam and a Michelson interferometer, which form an in-loop sensor for the proposed stabilization scheme. This sensing technique exploits the concept of a nondemolition measurement, since the power fluctuations are inferred by measuring the fluctuations in the phase observable of the auxiliary beam. This process results in higher in-loop signals for power fluctuations than what would be achieved by a direct detection, e.g. via the traditional scheme where a fraction of the laser power is picked off and sensed directly by a photodetector. Other advantages of this scheme are that the full beam power is preserved and available for further use, and that it enables the generation of a strong bright squeezed out-of-loop beam.

A brief introduction on laser power noise and the importance of power stabilization for interferometric gravitational wave detectors is given in Chap. 1. Here, an overview of power stabilization techniques reported in the state of the art is also presented. In sequence, an extensive theoretical investigation on the concept of the new power sensing scheme is presented in Chap. 2. In this investigation, different schemes in which power fluctuations are transferred to another observable of the light field, e.g. phase or polarization, are compared to each other, and the advantages of the radiation pressure transfer scheme are highlighted. Additionally, schemes to perform the readout of the power dependent phase/polarization are discussed. This chapter concludes that the most promising configuration to sense laser power fluctuations consists of a transfer scheme via radiation pressure on a suspended mirror, and a Michelson interferometer to readout the induced phase fluctuations in reflection of this mirror. The fundamental limits of this scheme are then analyzed in Chap. 3. In the beginning of this chapter, the two-photon formalism is introduced, which is then used to calculate the quantum noise limit of the radiation pressure scheme. Furthermore, the thermal noise limit and the total fundamental limit of the scheme are calculated for different parameters of suspended mirrors present in the state of the art. The analysis shows that a relative power noise below 10^{-9} $Hz^{-1/2}$ could be achieved at 10 Hz in the out-of-loop beam, which is of great interest for gravitational wave detectors. Remarkably, it also shows that a bright squeezed beam with a power of 4 W and up to 11 dB of squeezing might be achievable in the near future.

Based on the results of the theoretical investigations performed in Chaps. 2 and 3, a proof-of-principle experiment was performed, and the details of its design are described in Chap. 4. This experiment was realized with micro-oscillator mirrors with masses ranging from 25 to 250 ng, and fundamental resonance frequencies from 150 to 210 Hz. The experimental results are presented in Chap. 5, which is divided into three parts. The first part reports on the characterization of the noise sources of the interferometer containing the micro-oscillator mirror, which represent the sensing noise of the proposed stabilization scheme. In the second part, the use of the interferometer as a sensor for laser power fluctuations is demonstrated, and its experimental limitations are discussed. Finally, the first results of a power stabilization via radiation pressure are shown for different micro-oscillator mirrors, and for different transfer beam mean powers. Power stabilization in the frequency range from 1 Hz to 10 kHz was demonstrated. The results for the out-of-loop power stability are presented for different beam powers, and a relative power noise of 3.7×10^{-7} $Hz^{-1/2}$ was achieved at 250 Hz for 267 mW. The stability performance was limited by the structural thermal noise of the micro-oscillators, which was particularly high due to operation at room temperature. The results from the investigations conducted in this thesis are a promising step towards generation of a strong bright squeezed beam, and towards an improved stabilization scheme to be used in the future generation of gravitational wave detectors. A conclusion of this work and its future perspectives can be found lastly in Chap. 6.

Keywords Laser power stabilization · Bright squeezing · Optomechanics · Radiation pressure · Micro-oscillator · Gravitational wave detector

Hannover, Germany
August 2021

Marina Trad Nery

Acknowledgements

The time I spent in the Albert Einstein Institute during my Ph.D. was one of the best in my life. Not only have I grown tremendously as a scientist but also as a person. I am very grateful to work in such a friendly and collaborative environment, that made me feel home. For being a key piece in creating this outstanding institute, with an exceptional infrastructure, I would like to thank Karsten Danzmann.

My coming to Germany was only possible because of a project that promotes mobility of scientists around the world. Hence, I would like to acknowledge the European Union for funding the GraWIToN project under FP7-Marie Curie Actions (606176), which supported my research and training for the first 3 years of my Ph.D. I would also like to acknowledge Deutsche Forschungsgemeinschaft (EXC-2123 Quantum Frontiers 390837967) for funding this project.

My deepest gratitude goes to Benno Willke for being an incredible supervisor. Thank you for giving me the opportunity to work on such an interesting topic! Thank you also for all the thought and care you put in your advice, and for the valuable knowledge that you patiently shared with me. I had so much fun discussing physics with you, in particular the philosophical ones (which were many)! I always left your office feeling happy I chose to be a physicist. Finally, thank you for always giving me courage and bringing me back to the ground in my challenging moments.

I would like to thank all my amazing and dear colleagues in the Laser Group, particularly to Fabian, Nina, Li-Wei, Joscha, Vitus, Patrick, and Henning. Without your support and advice, this thesis would have not been the same! Thank you also to my office mates Kanioar, the Alps dual boy, and Jan, the philosopher. It was so good to share the office with you! Thank you for the food supply, and for being the ones with whom I could discuss anything, from physics to my personal dilemmas. My special thanks goes to Jasper (the mister many things) for the substantial contributions towards this experiment. Many were the challenges we had to overcome together to achieve the results shown in this thesis: from complicated ground loops, weird electronics, alignments, and things we still don't understand, to the big Telfon, bugs on the spectrum analyzer, and the last thing we would look for: a broken cable. You were always there! It was so good to have you to share many laughs and also some occasional frustrations in the lab.

A super big thank you for all the sharp comments and suggestions from the proof readers of this thesis: Jasper (for the nit picking!), Joscha (for the enthusiastic discussions on Chap. 2!), David, Sean, Sweta, Jan, Fabian, and Sina.

The radiation pressure experiment performed in this thesis was only possible due to a collaboration with the group of Thomas Corbitt, from the Louisiana State University. Thank you Thomas for all the knowledge you shared with us and for being so open in initiating this collaboration. I would like to thank also Stefan Danilishin, for all the interesting and long discussions about quantum mechanics, which were crucial for the calculations performed in Chap. 3.

A thank you to all the administrative staff of the institute, especially to Oksana for the excellent job in helping me integrate in Germany as a foreigner. Thank you also to the mechanics and electronics workshop, for building many components of this experiment, especially to Andreas for convincing me not to implement completely and utterly impossible ideas in the electronics of the setup, and for successful years in the lab without any explosions so far.

I also want to thank the members of the most diverse group I was ever part of: Sweta, Sonja, Neda, and Aparna. There was not a single time we were together that I did not feel happier than I was before! You taught me so much. A special thanks to Sonja for the wonderful friendship, and for being the closest feeling to a family that I had in Germany. I am so happy we met! A warm thank you also for the several generations of the Team Mensa members and its aggregates, especially to German, Karolina, David, Julia, Thomas, Lea, Daniel, Yele, Juliane, Stefan, Melanie, Vaishali, and Jonathan. I always look forward to our meetings and I keep many special memories and stories from our lunch times, parties, trips, and adventures. I will never forget this time we lived together!

For supporting the scientist in me from when I was a young child afraid of aliens, until I decided to move to another continent in pursue of my Ph.D., I would like to thank my parents. I want you to know that even with this far distance, I have always kept you very close to my heart. And finally, to my dear husband Vitor. Thank you for being crazy enough to jump on this adventure with me, and for being such a strong and positive person during these years. I have no words to express all the joy you brought into my life. You are surely the best partner I could have asked for!

Contents

Acronyms

AOM	Acousto-optic modulator
ASD	Amplitude spectral density
CCD	Charge-coupled device
DAN	4-(N, N-dimethylamino)-3-acetamidonitrobenzene
DAST	4'-dimethylamino-N-methyl-4-stilbazolium
GWD	Gravitational wave detector
HV	High voltage
KTP	$KTiOPO_4$
LIGO	Laser interferometer gravitational wave observatory
MI	Michelson interferometer
MI PD	Michelson interferometer photodetector
NPRO	Non-planar ring oscillator
OOL	Out-of-loop
OOL PD	Out-of-loop photodetector
PBS	Polarizing beamsplitter
PD	Photodetector
PM PD	Power monitor photodetector
PS PD	Power stabilization photodetector
PTFE	Polytetrafluorethylen
PTS	Polydiacetylene
PZT	Piezoelectric transducer
QCL	Quantum cascaded lasers
RPM	Relative power modulation
RPN	Relative power noise
RSN	Relative shot noise
SN	Shot noise
UGF	Unity gain frequency

List of Figures

Chapter 1
Introduction

Abstract This chapter presents an introduction to the field of laser power stabilization and its application to gravitational wave detectors. It starts with an introduction to gravitational waves and gravitational waves detectors. Then, the main coupling paths of power noise in gravitational wave detectors are discussed and the power noise requirement for the Advanced LIGO detector is presented. Finally, a mathematical description of power noise is given and an overview of the state of the art of power stabilization schemes for gravitational wave detectors is presented. In this overview, the limitations of current schemes are discussed and, lastly, the idea of the alternative scheme proposed in this thesis is introduced.

1.1 Motivation

Highly stable continuous-wave lasers are essential tools for many modern experiments, since laser noise is often a hindrance to their sensitivity. Examples of applications that demand a high stability in the laser power are high precision spectroscopy [1], cold atoms experiments [2], optomechanical experiments [3], and atom interferometry [4].

The most demanding requirement in power stability at low frequencies was set by ground based interferometric gravitational wave detectors (GWDs), where a maximum relative power noise of 2×10^{-9} Hz$^{-1/2}$ was required at 10 Hz. Because this value was approximately one order of magnitude below the residual relative power noise achieved in previous stabilization experiments [5], innovative techniques had to be developed in order to overcome prior limitations and fulfill the requirements for the current generation of GWDs. In the coming years, a generation of GWDs, approximately one order of magnitude more sensitive than the current one, is planned to be built. This increase in sensitivity will most likely demand an even higher power stability, and a relative power noise below 10^{-9} Hz$^{-1/2}$ might be required, which has not yet been demonstrated in the frequency regime relevant for GWDs. Hence, the main motivation for the work carried out in this thesis was to investigate alternative schemes that could satisfy more stringent power stability requirements and be implemented in the future generations of GWDs. The application of such new schemes is,

M. Trad Nery, *Laser Power Stabilization via Radiation Pressure*, Springer Theses,
https://doi.org/10.1007/978-3-030-95868-8_1

however, not limited to GWDs, and other high precision experiments, especially the ones involving optomechanics, can benefit from this work.

1.2 Gravitational Waves

Gravitational waves are ripples in the curvature of spacetime which are emitted by extremely energetic cosmological events, like colliding black holes and colliding neutron stars, amongst others. They propagate through spacetime at the vacuum speed of light and carry energy containing important information from their sources, such as mass, spin, and orbital parameters. Unlike electromagnetic waves, they propagate with almost no absorption and scattering. In addition to that, many sources of gravitational waves will not be sources of electromagnetic waves, which make GWDs a powerful and unique tool to study the Universe.

The first proposal of gravitational waves was made by Henri Poincaré in 1906 [6]. Their existence was subsequently predicted in 1916 by the renowned General Theory of Relativity of Albert Einstein [7]. The theory predicts that gravitational waves are emitted by spherically asymmetric accelerated masses, more specifically by a time varying quadrupole moment of a mass distribution. An example of a gravitational wave source is a binary system of two neutron stars orbiting around their common center of mass. The system dynamically changes the curvature of spacetime, and consequently emits gravitational waves that carry energy away from the system. As a result, the orbital separation of the system is constantly being reduced until the stars collide with each other. In fact, the first experimental proof, albeit indirect, of the existence of gravitational waves came from observations of the binary pulsar system PSR 1913+16, discovered in 1974 [8]. Russell A. Hulse and Joseph H. Taylor showed that the stars were getting closer to each other at precisely the rate predicted by general relativity, due to the emission of gravitational waves. A Nobel Prize in Physics was awarded in 1993, as a recognition of the major discovery of the pulsar system and its findings.

Gravitational waves are transversal waves, with oscillations that can be decomposed into two polarization components, called the plus $+$ and the cross $\times$ polarization. The effect of a gravitational wave in the metric of spacetime is transversal to its propagation direction. A ring of freely falling[1] test masses, with initial diameter L, is squeezed in one direction and equally stretched in the orthogonal direction, as a consequence of a gravitational wave propagating in the perpendicular direction to the ring plane. The strength of the effect of a gravitational wave is expressed by the strain h, defined by:

$$h = \frac{2\Delta L}{L} \,. \tag{1.1}$$

[1] Object that is under influence of only gravitational forces.

Due to the stiffness of the spacetime[2] and to the fact that the amplitude of a gravitational wave is inversely proportional to the distance from the source to the observer, even the strongest waves, emitted by violent cosmological events, will produce small effects when measured on Earth. Let us consider a simple example of a binary neutron star system, each with a mass of 1.4 times the mass of the Sun, and located in the Virgo Cluster, the nearest cluster of galaxies to the Earth. When the orbit of the stellar system has shrunk to a point where the individual stars are almost touching, it will emit gravitational waves that have a strain h on the order of 10^{-21} on Earth [11]. Such an impressively small strain is extremely hard to detect, and Einstein himself did not believe a detection would ever be possible. Nevertheless, on September 2015, 100 years after their prediction by general relativity, the first direct detection of gravitational waves, emitted by two merging black holes, was made by the two LIGO[3] detectors [12]. The detection was acknowledged as one of the biggest breakthroughs of our times, and the 2017 Nobel Prize in Physics was awarded to Rainer Weiss, Barry C. Barish, and Kip S. Thorne for their contributions to the field.

After the first detection, a new era for astronomy began, giving access to independently observe parts of our Universe and also to knowledge that was inaccessible via electromagnetic waves. From the first[4] and second[5] observing runs of the Advanced LIGO and Virgo[6] detectors, a total of 12 merger events, 11 from binary black holes and 1 from a binary neutron star, were detected [13]. From the binary black holes observations, tests on general relativity theory were performed [14], and the population properties of binary black holes, such as mass, spin, redshift, and merger rate density, could be inferred [15]. The merger of the two neutron stars [16] not only reassured the potential of GWDs, but also settled any doubts that the detected events had astrophysical origin. This was because the collision had an electromagnetic counter part which was observed by telescopes in almost all frequencies of the electromagnetic spectrum [17], validating the observation. This electromagnetic follow-up was only possible due to a narrow sky localization of the merger host galaxy, provided by the GWDs. The scientific outputs from this detection were many: a confirmation that neutron star mergers are a progenitor of short gamma ray bursts [18], a verification that gravitational waves travel at the speed of light, and more knowledge about the astrophysics of neutron stars [19, 20]. Another important outcome from the detections was an independent set of measurements of the Hubble constant using the luminosity distance obtained from the gravitational wave signals from the neutron stars (standard siren [20]), from the binary black holes (dark standard siren [21]), and

---

[2] Interesting fact: the corresponding Young's modulus of spacetime is 10^{20} times higher than the Young's modulus of steel [10]!

[3] LIGO stands for Laser Interferometer Gravitational-Wave Observatory, and it represents two large-scale interferometeric GWDs, one located in Hanford, and the other in Livingston in the United States.

[4] From the 12th of September 2015 to the 19th of January 2016.

[5] From the 30th of November 2016 to the 25th of August 2017.

[6] Virgo is a large-scale interferometric GWD located in Cascina, Italy.

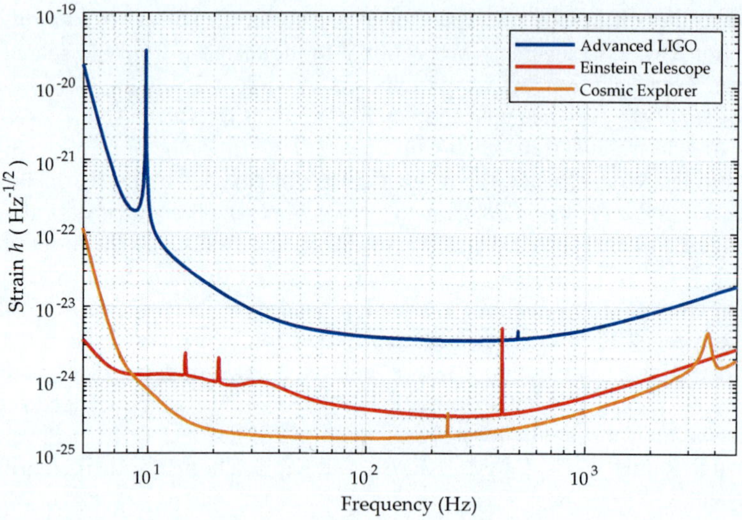

Fig. 1.1 Amplitude spectral density of the target strain sensitivity for a next generation of gravitational wave detectors, namely the Cosmic Explorer and the Einstein Telescope. The design sensitivity of the Advanced LIGO detector (current generation) is shown, as a reference, by the blue curve. The curves were plotted with data from [25]

from a combination of the multiple signals [22]. Recently, the third[7] observing run of Advanced LIGO and Virgo resulted in observations of more than 53 gravitational wave candidate signals, which are currently being analyzed. These include a candidate for a neutron star and black hole merger, a possible neutron star collision [23], and a novel signal from a merger of two black holes with unequal masses[8], where one black hole was approximately 3.6 times more massive than the other [24].

All observing runs provided more understanding about our Universe and future detectors, with an increased sensitivity, have a strong potential for many more findings. Research and development on a future generation of GWDs is currently very active [26, 27]. As shown in Fig. 1.1, the goal is to increase the strain sensitivity by more than one order of magnitude in the detection bandwidth, which will make possible to observe a larger volume of the Universe. The science for the next generation of ground based detectors includes a survey of primordial stellar mass black holes formed in the early Universe (redshifts of $z \approx 20$) [28], test of matter in extreme environments, and the study of phenomena which radiate weaker gravitational waves than compact binary systems, such as the core-collapse of supernovae [29] and continuous gravitational waves emitted by neutron stars [30]. Cosmology will also benefit from a possible detection of primordial gravitational waves [31], and from tests of dark energy and dark matter theories [32, 33]. Together with the launch of a space based

[7] From the 1st of April 2019 to the 23rd of March 2020.

[8] All previous detections were comprised of black holes of nearly equal masses.

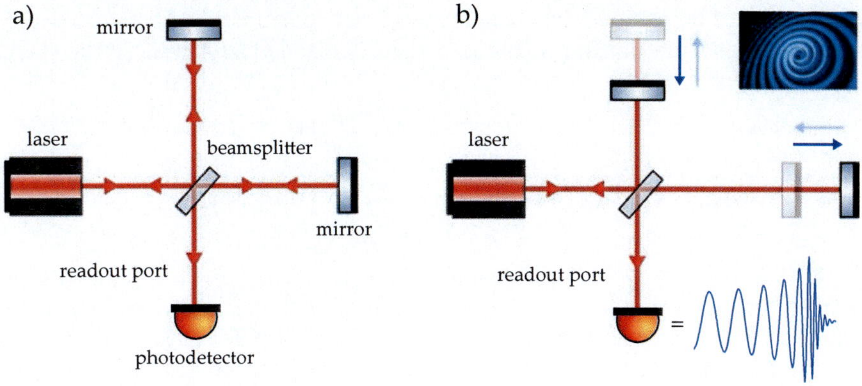

Fig. 1.2 Schematic of **a)** a Michelson interferometer and **b)** a Michelson interferometer under influence of a gravitational wave. Gravitational wave image credit: LIGO/T. Pyle

gravitational wave detector, the future promises exciting scientific explorations about our Universe, and possibly evidences for new physics, never anticipated before!

1.3 Interferometric Gravitational Wave Detectors

Currently, GWDs are based on a Michelson interferometer topology, as illustrated in Fig. 1.2a. The light coming from the laser source is divided in two orthogonal beams by a beamsplitter. The beams are back reflected by the interferometer end mirrors and are recombined at the beamsplitter, where interference occurs. As a result, part of the light will exit the interferometer at the readout port, and be detected by a photodetector,[9] and part of the light will be back reflected to the input of the interferometer. The detected power at the readout port is a function of the difference between the optical path lengths of the interferometer arms. Hence, by treating the interferometer end mirrors as freely falling test masses, a differential arm length change induced by a gravitational wave can be detected by measuring the power change at the readout port of the interferometer, as illustrated in Fig. 1.2b.

Previous table-top Michelson interferometer experiments were able to measure differential arm length changes of $\approx 10^{-14}$ m between 10 Hz and 10 kHz. The current generation of LIGO however, was designed to have a strain sensitivity smaller than $10^{-23}\,\mathrm{Hz}^{-1/2}$ (see Fig. 1.1). This means that the interferometer, which has a 4 km arm length, was required to measure differential arm length changes of 10^{-20} m, which is more than 5 orders of magnitude smaller than the charge diameter of a proton! To achieve this extremely high sensitivity, several noise sources like laser noise, seismic

[9] A photodetector is one of the most common devices to measure optical power and is comprised by a photodiode, which converts the detected power into an electric current, and a transimpedance amplifier circuit.

noise, thermal noise, and even quantum noise, had to be reduced to unprecedented levels. In the next section, the laser power stability required for Advanced LIGO will be introduced.

1.4 Power Noise in Gravitational Wave Detectors

Figure 1.3 shows the power noise requirements at the interferometer input, and, as a reference, the free running power noise of the 200 W laser system of Advanced LIGO. The requirements were calculated considering the different coupling paths of laser power noise into the interferometer readout port [34], explained hereby.

The direct coupling path for power noise, which set the requirement for frequencies between 2 and 10 kHz, is via the operational point in which the interferometer is locked. If the static differential arm lengths of the interferometer are adjusted such that destructive interference occurs in the field at the readout port, the interferometer is said to be locked at the dark fringe operational point. In this configuration, ideally no power is detected by the photodetector in the absence of a gravitational wave. Hence, the dark fringe is an optimal operational point in order to avoid direct power noise coupling at the readout port. However, to increase the strength of the signal caused by a gravitational wave, GWDs are currently operated slightly off the dark fringe, by means of a technique called DC readout [35]. As a result, light will intentionally exit the readout port of the interferometer, even in the absence of a

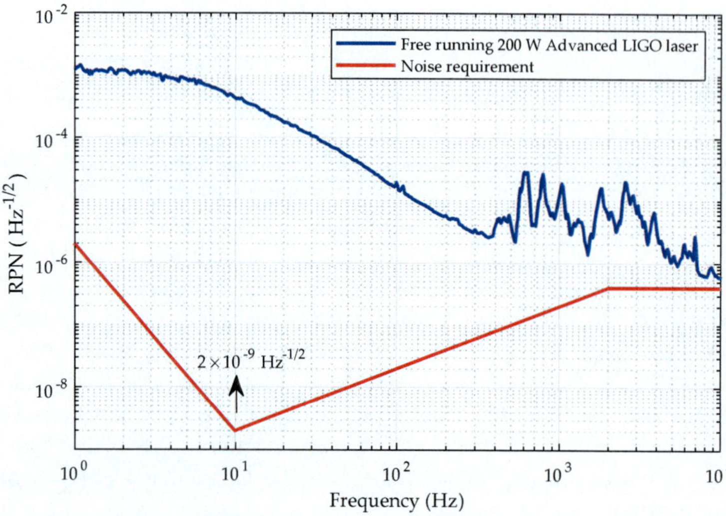

Fig. 1.3 Amplitude spectral density of the relative power noise (RPN) requirements at the interferometer input, calculated for the Advanced LIGO detector [34]. The blue curve shows, as a reference, the free running power noise of the 200 W laser system at LIGO

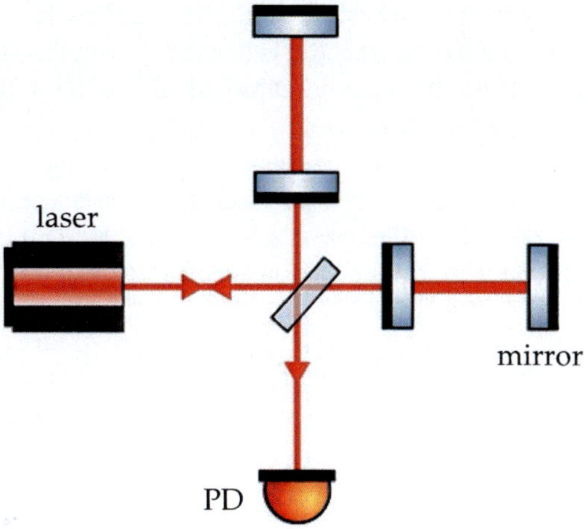

Fig. 1.4 Schematic of a Michelson interferometer with Fabry-Pérot arm cavities

gravitational wave. In this configuration, power noise at the input of the interferometer will couple directly as power noise at the interferometer readout, and can mask the signal of a gravitational wave.

In reality, GWDs are far more complex than a simple Michelson interferometer, and additional coupling paths for power noise are present. To amplify the signal of a gravitational wave, the power in each interferometer arm is increased by implementing a Fabry-Pérot cavity, as shown in Fig. 1.4. Due to technical reasons, an imbalance between the circulating power in the two cavities is unavoidable, which is assumed to be of approximately 1%. As a consequence, laser power noise will induce different radiation pressure forces on the cavity mirrors, which are suspended by multiple pendulums in order to reduce the effect of seismic noise on the gravitational wave measurement. This differential motion caused by radiation pressure will then couple to the interferometer readout port as displacement noise. This coupling path set the relative power noise requirements for frequencies below 2 kHz, including the most stringent value of 2×10^{-9} Hz$^{-1/2}$ at 10 Hz. For frequencies between 10 Hz and 2 kHz the requirement was calculated by imposing that the power noise coupling via radiation pressure should be 10 times smaller than the target strain sensitivity of the interferometer. For frequencies between 0.1 and 10 Hz, the requirement was calculated by imposing that the radiation pressure motion of the mirrors caused by the laser power noise should be smaller than their seismically induced motion.

Other power noise couplings resulting in less stringent requirements are also present, such as power noise coupling at the utility frequency and its harmonics, and power noise at the modulation/demodulation frequencies necessary to control several degrees of freedom of the interferometer.

The power stability requirements for the future generation of GWDs has not been yet calculated since it highly depends on their specific configuration, which is currently under discussion. However, it is expected that higher power stabilities will be demanded by the increase in their sensitivity. For example, the radiation pressure coupling path might be even more relevant in the future, since at 10 Hz a sensitivity increase by more than 2 orders of magnitude is planned, together with an increase of at least one order of magnitude in the intracavity power in the interferometer arm cavities [26, 27].

Hereafter, a brief review of the challenges and techniques to stabilize the laser power and meet the requirements of the current, and possibly of the future generation of GWDs, will be presented. But first, let us introduce laser power fluctuations and its sources.

1.5 Introduction to Laser Power Fluctuations

Mathematical Description

For simplicity, let us describe the laser light field by a monochromatic and linearly polarized plane wave. The amplitude fluctuations of its complex electric field $\tilde{E}$ can be described, to first order, by a Fourier series, i.e., by a sum of sinusoidal waves with amplitude m_Ω and Fourier angular frequency Ω:

$$\tilde{E}(t) = E e^{i\omega_0 t}\left(1 + \sum_\Omega m_\Omega \cos(\Omega t)\right), \tag{1.2}$$

where ω_0 is the laser optical angular frequency, and $0 \leq m_\Omega \leq 1$. For small amplitude fluctuations ($m_\Omega \ll 1$), the laser power P, which is proportional to $\left|\tilde{E}(t)\right|^2$, can be approximated to:

$$P(t) \approx \overline{P}\left(1 + \sum_\Omega 2m_\Omega \cos(\Omega t)\right) \equiv \overline{P} + \delta P(t), \quad [P] = \mathrm{W}, \tag{1.3}$$

where $\overline{P}$ is the mean optical power, averaged over a given time interval, and δP represents the absolute time dependent power fluctuations. This equation shows that the fluctuations δP are proportional to the mean laser power. One could then argue that in order to reduce the absolute power fluctuations it is only necessary to reduce the mean power. However, in most of the metrology experiments, the signal of the interested quantity to be measured scales with the mean laser power. Hence, it is of interest to have a laser with high power $\overline{P}$ and small fluctuations δP. The most common figure of merit to describe laser power fluctuations is the relative power noise (RPN), which is defined as:

$$\mathrm{RPN} = \frac{\delta P}{\overline{P}}. \tag{1.4}$$

In this thesis, noise will always be characterized in the frequency domain, which is a convention often adopted since noise is usually a result of a stochastic process. The fluctuations will then be represented by their amplitude spectral density (ASD), which expresses the strength of the fluctuation at a certain Fourier frequency Ω, per Hz, and will lead to the following units for the absolute and relative power fluctuations:

$$[\delta P\,(\Omega)] = \mathrm{W} \cdot \mathrm{Hz}^{-1/2} \quad \text{and} \quad [\mathrm{RPN}\,(\Omega)] = \mathrm{Hz}^{-1/2}\,. \tag{1.5}$$

Laser power fluctuations can be divided into two categories, according to their source: technical fluctuations and shot noise. Usually, solid state lasers are limited by technical fluctuations for Fourier frequencies up to several MHz (below the laser relaxation oscillation), and are shot noise limited at higher frequencies.

Technical Fluctuations

Technical power fluctuations can originate internally or externally to the laser source. Fluctuations at the laser source are caused by several phenomena such as noise in the laser pump current, thermal noise in the laser gain medium, length noise in the laser resonators, and laser relaxation oscillation. If the laser source involves amplifying stages, which are necessary to achieve high powers in GWDs, power noise can be introduced by these stages, in addition to the amplified power noise of the seed laser [5, 36].

Outside the source, the laser beam encounters several optical devices that can introduce power fluctuations. Beam jitter and length changes in an optical resonator, for example, will cause additional power fluctuations in its transmitted beam. Polarization dependent components, such as mirrors and polarizing beamsplitters (PBSs), will couple polarization fluctuations into power fluctuations. Additional power noise can also couple directly at photodetectors, which are common devices to measure the output optical signal in metrology experiments. For example, dust particles passing through the beam and scattered light from moving surfaces will cause fluctuations in the detected photocurrent. Beam jitter at the photodiode can also cause an apparent power noise via inhomogeneities in the responsivity of the photodiode. All these technical noise sources should be reduced as much as possible when designing an experiment in order to lower the power noise coupling and to relax the requirements in the laser power stability.

Shot Noise

Shot noise is a fundamental source of laser power fluctuations that is intrinsic to the discrete description of light as photons. Due to their discrete nature, the arrival rate of photons during a measurement interval, which is equivalent to the laser power, is statistically characterized by a Poisson distribution. For a large number of photons, which is usually the case in laser beams, this characterization can be approximated by a Gaussian distribution with a standard deviation equal to the square root of the mean number of photons per measurement interval [11]. The single sided ASD of the absolute laser shot noise (SN) is given by[10]:

$$\mathrm{SN} = \sqrt{2\hbar\omega_0 \overline{P}}\,, \quad [\mathrm{SN}] = \mathrm{W} \cdot \mathrm{Hz}^{-1/2}\,, \tag{1.6}$$

where $\hbar = h/2\pi$ is the reduced Planck constant. As expected, the absolute shot noise is proportional to the square root of the mean optical power, and its spectral density is frequency independent (white noise). The ASD of the relative shot noise (RSN) is then given by:

$$\mathrm{RSN} = \sqrt{\frac{2\hbar\omega_0}{\overline{P}}} \approx 2 \times 10^{-8}\ \mathrm{Hz}^{-1/2} \sqrt{\frac{1\ \mathrm{mW}}{\overline{P}} \cdot \frac{1064\ \mathrm{nm}}{\lambda_0}}\,. \tag{1.7}$$

Shot noise sets a lower limit to the laser power fluctuations that a classical laser beam can exhibit. Stabilization schemes employing non-classical states of light can surpass this limit and produce light with a sub-shot noise power stability, which is also known as amplitude squeezed light or, equivalently, as bright squeezed beam. Throughout this thesis, power stabilization schemes with the goal of stabilizing both technical and shot noise will be discussed.

1.6 Review of Power Stabilization Schemes

Stabilization schemes are usually divided into passive and active schemes. Passive schemes mostly implement filter devices, such as optical resonators to filter laser power noise in transmission of the resonator, and low pass filters to filter electronic noise in the laser electronics supply. Sophisticated passive schemes exploiting the Kerr effect are also possible, and it will be discussed in Sect. 2.2.2. Active stabilization schemes, on the other hand, implement a negative feedback control loop with a high gain to suppress power fluctuations sensed by an in-loop detector. Passive and active schemes can be combined and also cascaded in order to achieve better results not only in the noise suppression, but also in frequency range of the stabilization. The disadvantage of passive schemes implementing filters is that they only suppress fluctuations by a constant factor that depends on the filter properties [5]. So far, no passive scheme has been able to provide sufficient noise reduction in order to achieve a relative power noise on the order of $10^{-9}\ \mathrm{Hz}^{-1/2}$ at frequencies in the gravitational wave detection bandwidth. For this reason, this section will focus only on the review of active stabilization schemes.

[10] This equation will be derived in Sect. 3.1.1, using the two-photon formalism description of the light field.

1.6.1 Traditional Power Stabilization Scheme

The most common and simple active stabilization scheme is the traditional scheme, illustrated in Fig. 1.5. In this scheme, a fraction of the laser light is reflected by a beamsplitter and sensed by a photodetector, which is the in-loop sensor of this scheme. The signal from the photodetector is compared to a stable reference and the resulting error signal is then amplified by a controller and sent to a power actuator. The actuator modulates the main laser beam power such as to bring and maintain the error signal close to zero, thereby suppressing the power fluctuations sensed by the in-loop detector. The beam transmitted by the beamsplitter, called out-of-loop beam, is then stabilized and available for further use.

As in any active stabilization scheme, the achievable noise reduction in the out-of-loop beam depends on characteristics of the actuator, controller, and sensor [37]. An important property for the actuator, is a sufficiently large dynamic range in order to suppress the difference between the reference and the free running fluctuations in the desired stabilization frequency bandwidth. The controller, for example, needs to provide a large signal amplification without introducing instabilities around the required unity gain frequency. With careful design, the actuator and controller usually do not limit the performance of active power stabilization schemes, especially in the bandwidth below 10 kHz, which is the main interest for GWDs. The in-loop sensor, however, currently imposes the most important limitation in the traditional scheme. The limitation originates from sensing noise, which accounts for all non-deterministic noise sources that couple to the sensor, and therefore cannot be distinguished from the power fluctuations δP_0 of the main laser beam. By trying to correct for the sensing noise, the control loop will imprint an additional power modulation in the

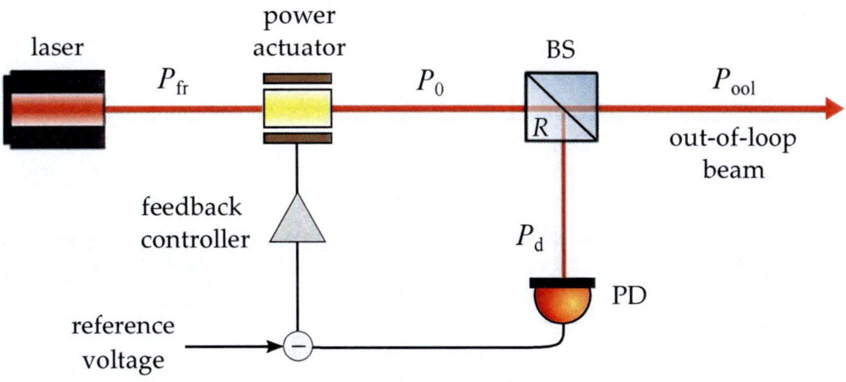

Fig. 1.5 Schematic of the traditional active power stabilization scheme. A fraction of the laser light is reflected by a beamsplitter (BS) with power reflectivity R, and sensed by a photodetector (PD). The signal from the photodetector is compared to a stable reference and the resulting error signal is amplified by a controller and sent to a power actuator. The actuator modulates the main laser beam power P_0 and suppresses the free running power fluctuations δP_{fr}. The out-of-loop beam, which is transmitted by the beamsplitter, has a stabilized power P_{ool} and is available for further use

main beam, that will set a lower limit for the power stability that can be achieved in the out-of-loop beam. Technical sensing noise can be caused, for example, by fluctuations in the reference voltage, electronic noise in the photodetector, beam jitter at the photodiode, and scattered light. To achieve a high power stability in the out-of-loop beam, it is crucial to reduce the technical sensing noise to a point where the scheme is approximately limited by shot noise, which is the fundamental limitation of the traditional scheme.

A rigorous understanding of why and how the laser shot noise imposes a limit to this scheme can only be achieved by taking into account the quantum nature of the light field and the beamsplitter, which will be further explained in Sect. 3.1.2. For now, let us consider that, due to the attenuation in the mean reflected and transmitted powers by the beamsplitter, the shot noise of the in-loop and of the out-of-loop beams are uncorrelated with the shot noise of the main beam with power $\overline{P}_0$. Hence, in the high loop gain regime, the free running power fluctuations δP_{fr} will be completely suppressed by the control loop, while the in-loop shot noise will be imprinted in the out-of-loop beam. In this situation, the absolute power fluctuations of the out-of-loop beam can be calculated as an uncorrelated sum of the in-loop shot noise $\mathrm{SN}_{\mathrm{il}}$, projected to the out-of-loop beam, and the out-of-loop shot noise $\mathrm{SN}_{\mathrm{ool}}$. This leads to:

$$\delta P_{\mathrm{ool}}^2 = \left(\frac{\mathrm{SN}_{\mathrm{il}} \cdot (1 - R)}{R} \right)^2 + \mathrm{SN}_{\mathrm{ool}}^2 \,. \tag{1.8}$$

By invoking Eq. 1.6, the absolute and relative power noise of the out-of-loop beam, in the high loop gain regime, can be written as:

$$\delta P_{\mathrm{ool}} = \sqrt{\frac{2\hbar\omega_0 \overline{P}_0 \cdot (1 - R)}{R}} = \sqrt{\frac{2\hbar\omega_0 \overline{P}_{\mathrm{ool}}}{R}} \,, \tag{1.9}$$

$$\mathrm{RPN}_{\mathrm{ool}} = \sqrt{\frac{2\hbar\omega_0}{\overline{P}_0 \cdot R\,(1 - R)}} = \sqrt{\frac{2\hbar\omega_0}{\overline{P}_{\mathrm{ool}} \cdot R}} \,. \tag{1.10}$$

Since $R < 1$, the resulting power noise in the out-of-loop beam will always be higher than its shot noise. For this reason, a sub-shot noise power stabilization is not possible with this configuration. Disregarding any other noise sources in the control loop, an arbitrarily high power stability in the beam transmitted by the power actuator could be achieved if the in-loop sensor would detect the full laser beam power P_0, i.e., if $R = 1$. This configuration is of course not practical since the laser power is completely wasted in the detection process and cannot be further used.

The value of $\mathrm{RPN}_{\mathrm{ool}}$ is minimized for a beamsplitter reflectivity of $R = 0.5$, which results in a relative power noise 2 times (or 6 dB) higher than the relative shot noise of the main beam P_0. Such reflectivity is undesirable in many experiments since only half of the initial power is available to the out-of-loop beam, and also due to technical difficulties in detecting high powers with a photodetector. Hence, let us consider that the detected power is much smaller than $\overline{P}_0$, i.e., that $R \ll 1$. In this case, the out-of-loop power noise can be approximated to:

$$\mathrm{RPN_{ool}} \approx \sqrt{\frac{2\hbar\omega_0}{\overline{P}_\mathrm{d}}} = 2 \times 10^{-9}\,\mathrm{Hz}^{-1/2}\sqrt{\frac{100\,\mathrm{mW}}{\overline{P}_\mathrm{d}}}, \tag{1.11}$$

for $\lambda_0 = 1064$ nm. This equation was derived assuming that the technical sensing noise is low enough such that the stabilization scheme is limited by shot noise. This is a realistic assumption in experiments that demand a relative power noise on the order of 10^{-8} $\mathrm{Hz}^{-1/2}$ at frequencies below a few hundred Hertz. However, a further reduction of technical sensing noise in experiments requiring higher power stability has proven to be much more challenging. For example, in [38], an $\mathrm{RPN_{ool}}$ of 5×10^{-9} $\mathrm{Hz}^{-1/2}$ was achieved at 10 Hz by detecting 110 mW with a photodiode quantum efficiency of 86%. This result was 1.8 times higher than what was expected from a shot noise limited performance, and this discrepancy was attributed to technical sensing noise, whose sources were not entirely understood. In order to tolerate technical noise, the Advanced LIGO power stabilization in-loop detector was required to detect approximately 200 mW to achieve an $\mathrm{RPN_{ool}}$ of 2×10^{-9} $\mathrm{Hz}^{-1/2}$ at 10 Hz. Since photodiodes with low noise and high efficiency performance typically use a small detector area, such high power imposes significant technical challenges due to the high thermal load on the photodetector [38, 39].

The solution adopted to overcome the technical limitation of detecting a high power in the in-loop detector was to split the in-loop beam power onto an array of 4 photodetectors, each detecting an individual power of approximately 50 mW [39]. The total in-loop shot noise of the array is the same as if all the power was detected by a single photodetector, since it is calculated as an uncorrelated sum of the shot noise of each detected beam. This scheme resulted in an $\mathrm{RPN_{ool}} = 2.4 \times 10^{-9}$ $\mathrm{Hz}^{-1/2}$ at 10 Hz, limited by electronic noise, and an $\mathrm{RPN_{ool}} = 1.8 \times 10^{-9}$ $\mathrm{Hz}^{-1/2}$ for frequencies above 15 Hz, which was closer than 1 dB to an in-loop shot noise limited performance. The implementation of the photodetector array was a contributor for the successful outcomes of the second observing run of Advanced LIGO [13], and fulfilled the given requirements for the current generation of GWDs. However, the sensitivity of this scheme can only be improved by further scaling the number of photodetectors and, due to the square-root dependence of the shot noise on the detected power, the technical effort required to achieve sensitivities close to $\mathrm{RPN_{ool}} = 10^{-10}$ $\mathrm{Hz}^{-1/2}$ ($\overline{P}_\mathrm{d} > 10$ W) is enormous. For this reason, alternative power stabilization schemes are desired for the future generation of GWDs.

1.6.2 Alternative Schemes

Optical AC Coupling

An alternative route investigated for laser power stabilization consisted on sensing power fluctuations in reflection of an optical cavity, like shown in Fig. 1.6a. This technique exploits the frequency dependency in the power transmitted and reflected by the cavity [40]. When the cavity is resonant with the laser beam, the mean power

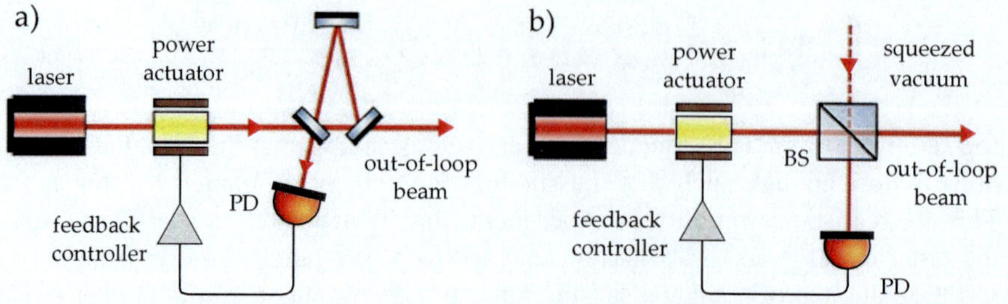

Fig. 1.6 Schematic of alternative active power stabilization schemes. **a** Optical AC coupling scheme: the in-loop photodetector is placed in reflection of an optical cavity, which is resonant with the laser beam. **b** Traditional scheme assisted by squeezing: amplitude squeezed vacuum is injected in the open port of the traditional scheme's beamsplitter, to reduce the in-loop and out-of-loop shot noise

of the laser is mainly transmitted by the cavity, while the power fluctuations with Fourier frequencies Ω higher than the cavity bandwidth are mostly reflected by the cavity (therefore the name of Optical AC Coupling scheme). Hence, this technique has the advantage of an increased sensitivity for power fluctuations without the need to increase the mean power on the photodiode, which is the main issue in the traditional scheme. This scheme was demonstrated for the first time with an optical cavity with a bandwidth of 35.7 kHz [40]. As a result, an $\mathrm{RPN}_{\mathrm{ool}}$ of $3.7 \times 10^{-9}\,\mathrm{Hz}^{-1/2}$ at frequencies around 200 kHz was achieved by detecting a power of approximately 0.3 mW. This relative power noise is almost one order of magnitude smaller than what would be achieved with the traditional scheme if the same power was detected by the in-loop photodetector. In a second generation of this experiment, a high finesse and 1 meter long optical cavity with a bandwidth of 4 kHz was implemented, and a stability of $8 \times 10^{-9}\,\mathrm{Hz}^{-1/2}$ was achieved at frequencies around 20 kHz [41].

One of the disadvantages of this technique is that, in order to perform the power sensing at low frequencies, a cavity with a bandwidth smaller than 10 Hz is required, which is extremely difficult to achieve in table top experiments. This is not an obstacle for long-baseline interferometers such as LIGO, which counts with a cavity bandwidth of around 0.6 Hz, formed by the power recycling resonator. However, investigations towards an implementation of the optical AC coupling technique in GWDs pointed to a high amount of additional noise in reflection of the power recycling resonator at low frequencies [42, 43]. As the detectors are highly complex instruments, the identification and reduction of these noise sources will not be a trivial task, which could be a hindrance in the performance of this scheme. In additional to that, the stabilization can only be tested and performed when the power recycling resonator is locked to resonance, which might interfere with other activities in the interferometer during commissioning time. From that perspective, a compact stabilization scheme that could be independently integrated in the laser system of the detector might be a better option. Hence, despite the successful implementations of this technique in table top experiments, it is not guaranteed that it will satisfy the future needs of GWDs.

Traditional Scheme Assisted by Squeezing

A technique which combines the traditional scheme with squeezed vacuum was also demonstrated as an alternative scheme for power stabilization [44]. In this scheme, a vacuum field squeezed in the amplitude quadrature was injected in the open port of the beamsplitter (see Fig. 1.6b) in order to reduce the in-loop and out-of-loop shot noise contributions. In this case, the out-of-loop relative power noise can be calculated by inserting the power spectral density S_{cc} of the amplitude quadrature of the vacuum field, into Eq. 1.11[11]:

$$\mathrm{RPN_{ool}} \approx \sqrt{\frac{2\hbar\omega_0 S_{cc}}{\overline{P}_d}}\,. \tag{1.12}$$

Because S_{cc} is smaller than $1\ \mathrm{s^{-1}\cdot Hz^{-1}}$ for an amplitude squeezed state, a sub-shot noise power stabilization is in principle possible with this technique, which is not the case for the traditional and optical AC coupling schemes [40]. However, a sub-shot noise performance for a high power in the out-of-loop beam is difficult to achieve. This can be seen by calculating the following condition:

$$\begin{aligned} \mathrm{RPN_{ool}} < \mathrm{RSN_{ool}} &\implies \frac{S_{cc}\cdot \overline{P}_0}{\overline{P}_d} < 1\,, \\ \text{with } \overline{P}_d &> 400\ \mathrm{mW}\cdot\frac{S_{cc}}{0.1}\cdot\frac{\overline{P}_0}{4\ \mathrm{W}}\,. \end{aligned} \tag{1.13}$$

An experimental demonstration of this scheme resulted in an $\mathrm{RPN_{ool}}$ of $2\times 10^{-8}\ \mathrm{Hz^{-1/2}}$ for frequencies between 5 and 80 kHz, by injecting a vacuum field squeezed by approximately 11 dB, and by detecting a power of only 106 μW. The achieved stability is 3 times (or 9.4 dB) higher than what would be achieved without injecting squeezed light at the beamsplitter.

One of the disadvantages of this technique is that it does not improve the tolerance to technical sensing noise which, as discussed in Sect. 1.6.1, was an important limitation at low frequencies in the traditional scheme.

Alternative Scheme Proposed in this Thesis

This thesis proposes an active stabilization scheme with an alternative concept for sensing laser power fluctuations. The concept consists on transferring the power fluctuations of the full laser beam, which will be called transfer beam, to the phase observable of a weak auxiliary beam, called sensing beam. The induced phase fluctuations in the sensing beam are then detected via a readout scheme, that results in power fluctuations δP_d that are sensed by an in-loop photodetector and fed back to the transfer beam power. As a result from this transfer, the detected power fluctuations in the sensing beam are amplified with respect to the transfer beam fluctuations δP_0,

---

[11] For a non squeezed vacuum field, $S_{cc} = 1\ \mathrm{s^{-1}\cdot Hz^{-1}}$. See Sect. 3.1.2 for the complete mathematical description of this scheme.

i.e., $\delta P_{\mathrm{d}} > \delta P_0$. Hence, this scheme provides a higher signal in the in-loop detector than what would be achieved by directly detecting the transfer beam power fluctuations, as in the optical AC coupling and the traditional scheme. In addition to that, the full optical power of the transfer beam is preserved and available for use. Finally, the proposed scheme has an additional advantage of a sub-shot noise stabilization performance in the out-of-loop beam which, in opposite to the traditional scheme assisted by squeezing, provides a stability proportional to the (transfer) beam power.

Chapter 2
Sensing Laser Power Fluctuations Via an Alternative Observable of the Light Field

Abstract Based on the limitations of the current power stabilization schemes discussed in Chap. 1, this chapter proposes an active scheme with an alternative method to sense power fluctuations. In the proposed method, the power fluctuations of the full laser beam induce fluctuations in another observable of the light field, such as the phase or the polarization. By measuring the new observable, the power fluctuations can be inferred. This sensing technique is similar to a nondemolition measurement since the observable of interest, in this case the laser amplitude, is nearly unperturbed during the measurement. An extensive theoretical investigation on possible schemes in which power fluctuations are transferred to another observable of the light field is given. The transfer scheme via radiation pressure is selected as the most promising one and its advantages are highlighted. Additionally, schemes to perform the readout of the power dependent phase/polarization are discussed.

Gravitational wave detectors, as other high precision metrology experiments, require a high laser power stability that is currently achieved using the traditional power stabilization scheme. As discussed in Chap. 1, the traditional scheme is limited by the high amount of power that needs to be detected by the in-loop photodetector in order to overcome the shot noise of its detected power and technical sensing noise. Different routes were adopted to circumvent this limitation such as substituting the single diode for an array of photodiodes [39], reducing shot noise by injection of squeezed vacuum [44], and sensing power fluctuations in reflection of an optical cavity (optical AC coupling) [40]. This chapter proposes an active power stabilization scheme with an alternative method to sense power fluctuations. In the proposed method, the power fluctuations δP_0 of the full laser beam induces fluctuations in another observable of the light field, such as the phase or the polarization. By measuring the new observable, the power fluctuations δP_0 can be inferred. This sensing technique is similar to a nondemolition measurement [45] since the observable of interest, in this case the laser amplitude, is nearly unperturbed during the measurement. One of the advantages of this method is that the readout of the new observable can be chosen such that the mean laser power is not attenuated during the sensing process, which preserves the power in the out-of-loop beam. As it will be shown, this scheme has the potential of

M. Trad Nery, *Laser Power Stabilization via Radiation Pressure*, Springer Theses,
https://doi.org/10.1007/978-3-030-95868-8_2

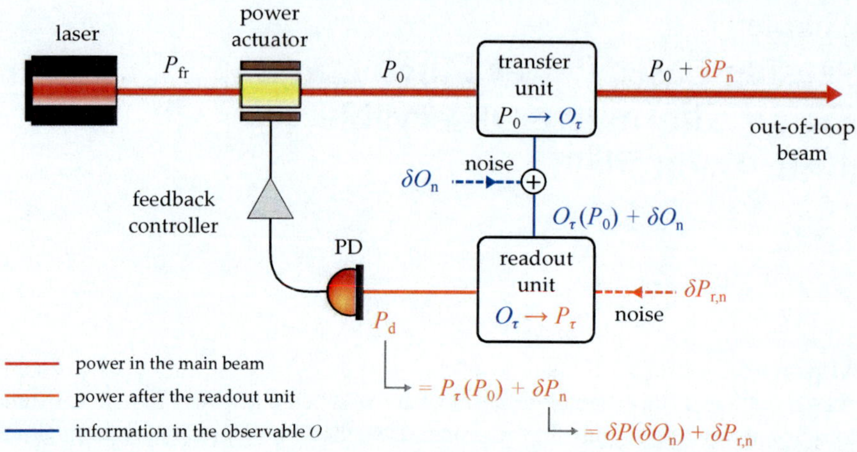

Fig. 2.1 Schematic of the concept of the proposed power stabilization scheme. A transfer unit induces a power dependence in an alternative observable O of the light field. The information in the observable, represented by the blue line, is converted back to power by a readout unit. The power at the output of the readout unit, which depends on P_0, is detected by a photodetector and fed back to a power actuator, which stabilizes the free running power P_{fr}. The figure also shows noise coupling at the transfer and readout units, that will ultimately couple as a power noise δP_n at the photodetector and will be imprinted in the out-of-loop beam

a power stabilization below the shot noise (bright squeezing) of the full beam power, which is not possible with the traditional and optical AC coupling schemes.

The general concept of the scheme is illustrated in Fig. 2.1. In a transfer unit, the power modulations[1] δP_0 from the full laser beam, induces modulations δO_τ in an alternative observable O of a light field, which can be the same field as the main beam (self modulation), or the field of an auxiliary beam (cross modulation). Typically, the new observable cannot be detected directly. Hence, a readout unit converts the modulations in the new observable (represented by the blue line) back to optical power modulations δP_τ, which is a function of δP_0. The power at the output of the readout unit is detected by a photodetector and fed back to the power actuator, which stabilizes the free running power modulations δP_{fr}. It is important to note that, ideally, the transfer unit does not alter the power P_0.

The total sensing scheme, comprised by the transfer and readout units, can be characterized by the following power transfer coefficient:

$$\eta = \frac{\delta P_\tau}{\delta P_0}, \tag{2.1}$$

[1] Without loss of generality, from now on this chapter will make reference to deterministic power modulations instead of random power fluctuations. This is to avoid confusion between the power fluctuations to be stabilized, from additional power noise sources in the scheme.

which expresses the transfer function between power modulations from the full beam, to power modulations at the photodetector. The figure also shows noise coupling via the transfer ($\delta O_{\rm n}$) and readout units ($\delta P_{\rm r,n}$), that cannot be distinguished from the induced modulations δO_τ, and will ultimately couple as power noise $\delta P_{\rm n}$ at the photodetector. As in any active stabilization scheme, the power noise $\delta P_{\rm n}$ will be imprinted in the out-of-loop beam by the feedback control loop, and set a limit to the power stability that can be achieved with this scheme.

Let us now compare the proposed scheme with the traditional and optical AC coupling schemes. For this comparison, let us first consider that the main beam power $\overline{P}_0$ and the detected power $\overline{P}_{\rm d}$ is the same for all schemes, and that $\delta P_{\rm n}$ has only the contribution of the shot noise of the detected beam, i.e., $\delta P_{\rm n} = {\rm SN_{il}}$. In the traditional scheme, the power modulations δP_0 are attenuated by the beamsplitter by a factor $\eta = R \ll 1$, considering that the main laser power is much larger than the detected power. For the optical AC coupling scheme, $\eta \approx 1$ for power modulations at frequencies higher than the cavity linewidth. In the scheme proposed in this thesis however, the transfer and readout units can be chosen such that $\eta \gg 1$. As a result, a higher signal-to-noise ratio $\delta P_\tau/{\rm SN_{il}}$ can be achieved in the in-loop detector with the proposed scheme, in comparison with the traditional and optical AC coupling schemes. However, a fair comparison between different stabilization schemes can only be made when all noise sources coupling to the in-loop and out-of-loop beams are taken into account. This analysis will be made in Chap. 3 for the most promising scheme for the transfer and readout units. The selection criteria for this scheme will be presented in the next sections of this chapter, where different schemes possibilities for the phase and the polarization observables are analyzed.

2.1 Phase Transfer Schemes

The complex electric field $\tilde{E}_\tau$ in transmission of a phase transfer scheme, when considering a linearly-polarized and quasi-monochromatic plane wave propagating in vacuum, can be written as:

$$\tilde{E}_\tau(t) = \tilde{E}_0(t)\, e^{i\phi_\tau(t)}\,, \tag{2.2}$$

where $\phi_\tau(t)$ is the power dependent part of the phase accumulated by the field in the transfer scheme, and $\tilde{E}_0(t)$ is the complex electric field at the input of the transfer scheme. It was assumed that the input amplitude and its modulations, as well as the polarization state of the field, are not altered by the phase transfer scheme. The power independent phase acquired by the beam through the transfer scheme was assumed to be constant, and therefore it was omitted since it is not relevant for sensing power modulations, and can be corrected by a compensating plate. It is important to keep in mind though, that fluctuations in this term caused by noise sources coupling to the transfer scheme have to be taking into account in the noise analysis of the scheme.

Assuming that the phase transfer is a linear function of the input power P_0, which should hold true for small power modulations, the induced phase can be written as:

$$\phi_\tau(t) = \tau \cdot P_0(t)\,, \tag{2.3}$$

where τ will be called phase transfer coefficient and is given in units of rad $\cdot$ W^{-1}. Since this work focuses on sensing power fluctuations of stable lasers in the continuous wave regime, values for δP_0 on the order of 10^{-6} W or smaller are expected, and the linear relation of Eq. 2.3 can always be justified.

Let us now estimate the required value for τ in order to detect a relative power modulation (RPM) of 10^{-9}. This value depends on the minimum phase modulation $\delta\phi_{\min}$ that can be detected by the readout scheme, limited by technical or fundamental noise sources. For this estimation it will be assumed that the readout scheme is sensitive enough to detect phase changes of 10^{-6} rad at 10 Hz, which was already demonstrated in the state of the art [46]. In this case, the following condition needs to be satisfied:

$$\tau \geq 10^3\ \mathrm{rad} \cdot \mathrm{W}^{-1} \left(\frac{\delta\phi_{\min}}{10^{-6}\ \mathrm{rad}} \cdot \frac{10^{-9}}{\mathrm{RPM}} \cdot \frac{1\ \mathrm{W}}{\overline{P}_0} \right)\,. \tag{2.4}$$

The frequency of 10 Hz was chosen for this estimation because it is the frequency of the most stringent requirement for power stability in Advanced LIGO (see Chap. 1), which is the main motivation of this work. The reader should keep in mind that at higher frequencies the technical noise sources in the readout schemes are usually smaller and the constraint on the transfer coefficient can be relaxed (lower values for $\delta\phi_{\min}$). In the next sections, phase transfer coefficients of different schemes are calculated and compared to each other.

2.1.1 Optical Kerr Effect

The optical Kerr effect is a third order nonlinear process in which an electromagnetic wave propagating in a nonlinear medium induces an intensity dependence in the medium's refractive index. As a result, the wave traveling through the medium will acquire an intensity dependent phase. The dependence of the refractive index on the intensity is introduced by the electric field of the wave which changes the orientation of the molecules inside the medium and induces an anisotropy in the medium's polarization, i.e., in the dipole moment per unit volume. Let us consider a monochromatic and linearly-polarized plane wave propagating through an isotropic medium with inversion symmetry ($\chi^{(2)} = 0$). Then, the medium's complex polarization $\tilde{\mathrm{P}}(t)$, up to the third order, is given by [47]:

$$\tilde{\mathrm{P}}(t) = \epsilon_0 \left(\chi^{(1)} \tilde{E}_0(t) + \chi^{(3)} \tilde{E}_0^3(t) \right)$$
$$\simeq \underbrace{\epsilon_0 \left(\chi^{(1)} + \frac{3}{4} \chi^{(3)} E_0^2 \right)}_{\chi} \tilde{E}_0(t) \,, \tag{2.5}$$

where ϵ_0 is the permittivity of free space. The constants χ, $\chi^{(1)}$, and $\chi^{(3)}$ are the total, first and third-order nonlinear optical susceptibilities, respectively, which represent the degree of polarization of the medium in response to the applied electric field. Polarization at frequencies $3\omega_0$, leading to third-harmonic generation, also occurs in the medium. They however, will not be taken into account in this thesis due to their small amplitude in comparison with the other contributions.

By definition, the total refractive index of the medium is given by:

$$n = (1 + \chi)^{1/2} \simeq n_0 \left(1 + \frac{3\chi^{(3)} E_0^2}{8 n_0^2} \right) , \tag{2.6}$$

where $n_0 = \left(1 + \chi^{(1)}\right)^{1/2}$ is the linear (low intensity) refractive index. This equation can be re-written as a function of the intensity of the incident light $I_0 = c n_0 \epsilon_0 E_0^2/2$, as:

$$n = n_0 + n_2 I_0 \,, \tag{2.7}$$

where c is the vacuum speed of light. The coefficient n_2 is the nonlinear refractive index, which expresses the strength of the optical nonlinearity and is given by:

$$n_2 = \frac{3\chi^{(3)}}{4 n_0^2 \epsilon_0 c} \,. \tag{2.8}$$

The total phase accumulated by the field after an interaction length L with the medium is $2\pi n L/\lambda_0$, which is a sum of a constant and intensity independent phase given by $2\pi n_0 L/\lambda_0$, and an intensity dependent phase ϕ_τ given by:

$$\phi_\tau(t) = \frac{2\pi \, n_2 L \, I_0(t)}{\lambda_0} \,. \tag{2.9}$$

This equation is valid for sampling times much smaller than the interaction time L/c. The phase transfer coefficient of this scheme is then:

$$\tau = \frac{2\pi \, n_2 L}{\lambda_0 \, \mathcal{A}} \,, \tag{2.10}$$

where $\mathcal{A} = \pi w_0^2$ is the effective cross-section area of the beam, with w_0 being the beam radius. Note that since the calculation was performed assuming a plane wave, the beam area is constant throughout the interaction length. It will be also assumed

Table 2.1 Phase transfer coefficients τ for the optical Kerr effect transfer scheme, calculated from Eq. 2.10. A beam radius of $w_0 = 20\,\mu m$ was considered

Material	n_2 $(cm^2 \cdot W^{-1})$	L (mm)	λ_0 (nm)	τ $(rad \cdot W^{-1})$
KTP [51]	2.4×10^{-15}	70	1064	7.9×10^{-5}
GaAs [52]	3×10^{-13}	1.2	1064	1.7×10^{-4}
PTS [53]	1×10^{-11}	0.2	1064	9.3×10^{-4}
InGaAs/InAlAs QCL [54]	8×10^{-9}	3	4770	2.5

throughout this thesis that the area $\mathcal{A}$ is not varying with time, and therefore modulations in the intensity are caused only by modulations in the power.

Since the third order susceptibility $\chi^{(3)}$ is usually on the order of 10^{-24} m^2/V^2, small phase transfer coefficients for this scheme are expected, as shown in Table 2.1. The crystal $KTiOPO_4$ (KTP) is a widely used material in nonlinear optics experiments, but it exhibits a very low third order susceptibility. Polymers, like polydiacetylene (PTS) for example, have much larger nonlinearities. They are, however, limited by the optical quality and the manufacturing process, which restricts the sample length to only 0.2 mm. With the phase transfer coefficient calculated for PTS, a power higher than 10^6 W would be needed to fulfill the constrain from Eq. 2.4. Such a high power in the continuous wave regime, which is of interest in this work, can trigger unwanted nonlinear effects and might also exceed the material's damage threshold. Several experiments in the literature compensate for the small transfer coefficient by using pulsed lasers, which have higher intensity, and also have the additional advantage of a lower damage threshold in the nonlinear materials.

The nonlinear refractive index of several other common materials can be found in [48]. More exotic options such as quantum cascaded lasers (QCL), liquid crystals ($n_2 = 10^{-3}$ $cm^2 \cdot W^{-1}$ [49]), quantum dots, quantum wells or cold atoms ($n_2 = 0.1$ $cm^2 \cdot W^{-1}$ [50]), can exhibit extraordinary high nonlinearities. However, they are not suited to the purposes of this thesis due to complexity, size, or transparency for laser wavelengths implemented in GWDs (1064 nm or 1550 nm).

An option to increase the phase transfer coefficient is to increase the interaction length L (see Eq. 2.10) and reduce the beam area $\mathcal{A}$, which can be done by exploiting the Kerr effect in optical fibers for example. A quantum nondemolition measurement in the MHz regime was demonstrated in [55], with a 114 m single-mode optical fiber cooled to 2 K. The experiment, however, exhibited a low signal-to-noise ratio due to phase noise, scattered light, and photon-phonon coupling. Another option would be to use fiber amplifiers, which are doped with rare earth elements, such as Erbium and Ytterbium, that can present high nonlinearities (order of 10^{-11} $cm^2 \cdot W^{-1}$ [56]) at the wavelength of the pump light (980 nm), and also be produced in lengths on the order of 20 m. Experiments show, however, that the nonlinear index of refraction n_2 changes in the presence of the pump power with a nonlinear behavior [57].

2.1.2 *Cascaded Kerr Effect*

There is an interesting option to mimic the Kerr effect by exploiting the second order susceptibility $\chi^{(2)}$ of a material, which is usually on the order of 10^{-12} m $\cdot$ V^{-1} and therefore much larger than $\chi^{(3)}$. The process is called cascaded Kerr effect and it can be understood in two steps, as shown in Fig. 2.2. In the first step, a fundamental wave with optical frequency ω_0 is partially up-converted to a second harmonic wave with optical frequency $2\,\omega_0$ via sum-frequency generation. In the second step, the second harmonic wave is down-converted via a difference-frequency generation between the second harmonic and the unconverted fundamental wave. As a result, a reconverted fundamental wave is generated which then interferes with the unconverted fundamental wave.

As a consequence of dispersion, the second harmonic wave acquires a phase shift with respect to the fundamental wave, and therefore the reconverted fundamental will have a phase shift of ϕ with respect to the unconverted wave. If the conversion process is phase matched, i.e., $\Delta k = 2k_0 - k_2 = 0$,[2] then the phase difference ϕ is equal to π. This will lead to depletion of the fundamental wave, as expected from energy conservation. In this case, the second harmonic power at the output of the medium is maximized and the fundamental wave does not accumulate any intensity dependent phase. If the process is not phase-matched, i.e., $\Delta k \neq 0$, then the total fundamental wave will suffer a phase shift. Because the up and the down-conversion steps are intensity dependent, the phase shift acquired by the fundamental wave will also depend on the intensity. The description of the cascaded effect in two steps is purely pedagogical, since in reality power is continuously exchanged between the fundamental and second harmonic waves throughout the interaction length. Hence, at the output of the medium, the fundamental wave will accumulate a net and intensity dependent phase ϕ_τ, mimicking the optical Kerr effect.

2.1.2.1 Cascaded Kerr Effect Type I

Let us consider that the input polarization of the fundamental wave is aligned to the ordinary axis of the nonlinear medium, and that the generated second harmonic wave has a polarization aligned to the extraordinary axis, which is known as type I condition for a positive uniaxial crystal [47]. The coupled equations for the complex electric fields $\tilde{E}_0$ and $\tilde{E}_2$ of the fundamental and second harmonic waves can be obtained by substituting the medium's polarization $\tilde{\mathrm{P}}$ into Maxwell's equations. In the absence of loss, and considering linearly-polarized and plane waves propagating along the z direction, the coupled equations are [58]:

---

[2] k_0 and k_2 are the wave vectors of the fundamental and second harmonic waves.

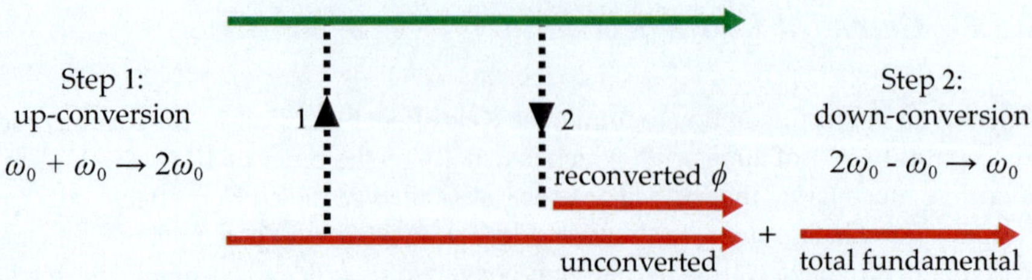

Fig. 2.2 Illustration of the cascaded Kerr effect type I. In the first step, the fundamental wave (red arrow) with optical frequency ω_0 is up-converted to the second harmonic wave (green arrow), with frequency $2\omega_0$. In the second step, the second harmonic wave is down-converted to a fundamental wave with a phase shift of ϕ, that interferes with the unconverted fundamental wave

$$\begin{aligned}\frac{\partial \tilde{E}_0}{\partial z} &= \frac{-i\omega_0\, d_{\text{eff}}}{n_{\omega_0} c}\, \tilde{E}_0^* \tilde{E}_2\, e^{-i\Delta k z}\,,\\ \frac{\partial \tilde{E}_2}{\partial z} &= \frac{-i\omega_0\, d_{\text{eff}}}{n_{2\omega_0} c}\, \tilde{E}_0^2\, e^{i\Delta k z}\,,\end{aligned} \tag{2.11}$$

where $d_{\text{eff}} = \chi^{(2)}/2$, and the frequency dependence of $\chi^{(2)}$ was neglected by assuming that all optical wavelengths are far from the material resonances. Note that the steps in the cascading process are both ruled by the same phase mismatch parameter Δk. These equations can be re-written in terms of the amplitude $\tilde{A}_\text{j} = \sqrt{\epsilon_0 n_\text{j} c/2}\, \tilde{E}_\text{j}$, such that $\left|\tilde{A}_\text{j}\right|^2$ expresses the field intensity in $\text{W} \cdot \text{m}^{-2}$:

$$\frac{\partial \tilde{A}_0}{\partial \varsigma} = -i\sigma\ \tilde{A}_0^* \tilde{A}_2 e^{-i\Delta k L \varsigma}\,, \tag{2.12}$$

$$\frac{\partial \tilde{A}_2}{\partial \varsigma} = -i\sigma\ \tilde{A}_0^2 e^{i\Delta k L \varsigma}\,, \tag{2.13}$$

where $\varsigma = z/L$, and $\sigma = L\omega_0 d_{\text{eff}} \sqrt{2/(c^3 \epsilon_0 n_{\omega_0}^2 n_{2\omega_0})}$ is the common nonlinear coupling coefficient, with $[\sigma^2] = \text{m}^2 \cdot \text{W}^{-1}$.

No-Depletion Regime

Let us assume that the conversion from the fundamental to the second harmonic field is small such that $\left|\tilde{A}_0(\varsigma)\right|$ remains constant during the cascaded process. Then, by integrating Eq. 2.13, the intensity of the second harmonic wave after propagating a length L is obtained:

$$I_2(L) = \sigma^2\, I_0^2\, \text{sinc}^2\left(\frac{\Delta k L}{2}\right)\,. \tag{2.14}$$

For $\Delta k \neq 0$, this equation expresses that the second harmonic intensity continuously grows inside the medium up to the coherence length, defined as $L_c = \pi/\Delta k$, which means that step 1 of the cascading process is dominating. After propagating a distance L_c, the phase difference of π between the waves forces the second harmonic wave to effectively down-convert to the fundamental wave. The fundamental field will be fully restored with a net phase shift of ϕ_τ after propagating a distance of two coherence lengths.

Assuming $\left|\tilde{A}_0(\varsigma)\right|$ constant, Eq. 2.12 has a solution of $\tilde{A}_0(L) = \tilde{A}_0(0)e^{-i\phi_\tau(L)}$ with [59]:

$$\phi_\tau(t) = \frac{\sigma^2 I_0(t)}{\Delta k L}, \tag{2.15}$$

valid for $\Delta k L \gg \sigma^2$. In analogy to n_2 in Eq. 2.9, an effective nonlinear index n_2^{eff} can be obtained:

$$n_2^{\text{eff}} = \frac{4\pi d_{\text{eff}}^2 L}{\lambda_0 \epsilon_0 c n_{2\omega_0} n_{\omega_0}^2} \cdot \frac{1}{\Delta k L}. \tag{2.16}$$

It is important to note that n_2^{eff} is a non-local property of the material since it is induced by the propagation of the waves and, unlike the Kerr effect, a cross phase modulation between two non-interacting waves is not possible.

Table 2.2 shows values of n_2^{eff} and its corresponding transfer coefficients τ, for the most common materials used for cascaded Kerr effect. For the first three rows, the values of n_2^{eff} were taken from experimental observations reported in the literature. For the last two rows, n_2^{eff} was calculated with the theoretical coefficient d_{eff} and for $\Delta k L = 2\pi$. This phase matching condition was chosen since it corresponds to a minimum of second harmonic power at the output of the medium, which is desired for the power stabilization scheme to avoid power loss in the fundamental beam. For the KTP crystal, an increase of 1 order of magnitude in the nonlinear index n_2^{eff} was experimentally demonstrated in comparison with the optical Kerr effect (displayed in the first row). In theory, an increase of up to 4 orders of magnitude is expected (displayed in the 4th row) when using a crystal with a length of 70 mm.

Table 2.2 Phase transfer coefficients τ for the cascaded Kerr effect type I transfer scheme in the low intensity regime, calculated from Eq. 2.10. The values of n_2^{eff} displayed in the 4th and 5th rows were calculated from Eq. 2.16, considering $\Delta k L = 2\pi$. A beam radius w_0 of 20 µm was considered

Material	L (mm)	λ_0 (nm)	n_2^{eff} $(\text{cm}^2 \cdot \text{W}^{-1})$	τ $(\text{rad} \cdot \text{W}^{-1})$
KTP [59, 63]	1	1064	1.3×10^{-14}	6×10^{-6}
DAN [64]	0.77	1064	2.5×10^{-13}	9×10^{-5}
DAST [63]	1	1540	3×10^{-11}	0.01
$d_{\text{eff}} = 3\,\text{pm} \cdot \text{V}^{-1}$ (KTP [48])	70	1064	9×10^{-13}	0.03
$d_{\text{eff}} = 290\,\text{pm} \cdot \text{V}^{-1}$ (DAST [61])	1	1540	8×10^{-11}	0.03

Second harmonic generation using a 30 W continuous-wave laser with 1064 nm and a 19 mm long periodically poled KTP crystal ($d_{\text{eff}} = 9\,\text{pm}\cdot\text{V}^{-1}$) was reported in [60]. With these parameters, an RPM of 10^{-9} would result in a phase change $\delta\phi_\tau$ of 6×10^{-10} rad, which is challenging to detect. Additionally, with this high power the deviation from the low intensity regime starts to be significant and smaller phase changes should be expected. Other common materials such as periodic poled lithium niobate (PPLN), or beta barium borate (BBO) exhibit similar properties which also makes them unsuitable for the purpose of this thesis. The table also shows the coefficient obtained for the organic material 4'-dimethylamino-N-methyl-4-stilbazolium (DAST), that can have a large nonlinearity of $d_{\text{eff}} = 290\,\text{pm}\cdot\text{V}^{-1}$ at 1540 nm [61] and is commercially available in lengths of up to 1 mm [62]. This material, however, presents strong absorption for the second harmonic wavelength (below 1 μm), which reduces the conversion efficiency of the cascaded process, and for this reason the measured nonlinear coefficient n_2^{eff} (displayed in the 3rd row) is smaller than the theoretical value (displayed in the 5th row).

High Intensity Regime

In the high intensity regime, there is significant conversion of the fundamental wave power into the second harmonic wave and therefore $\left|\tilde{A}_0(\varsigma)\right|$ cannot be treated as a constant inside the medium anymore. In this case, the coupled equations can be solved numerically or analytically, in terms of the Jacobi elliptic functions. By writing the complex fields as $\tilde{A}_{\text{j}}(\varsigma) = a_{\text{j}}(\varsigma)e^{-i\phi_{\text{j}}(\varsigma)}$, the coupled equations can be written separately as a function of the amplitude a_{j} and phase ϕ_{j}:

$$\begin{aligned}
\frac{\partial a_0}{\partial\varsigma} &= \sigma\, a_0 a_2 \sin(\Phi - \Delta k L\varsigma)\,,\\
\frac{\partial a_2}{\partial\varsigma} &= -\sigma\, a_0^2 \sin(\Phi - \Delta k L\varsigma)\,,\\
\frac{\partial \phi_0}{\partial\varsigma} &= -\sigma\, a_2 \cos(\Phi - \Delta k L\varsigma)\,,\\
\frac{\partial \phi_2}{\partial\varsigma} &= -\sigma\, \frac{a_0^2}{a_2} \cos(\Phi - \Delta k L\varsigma)\,,
\end{aligned} \tag{2.17}$$

where $\Phi(\varsigma) = \phi_2(\varsigma) - 2\phi_0(\varsigma)$. The numerical solutions for these equations are plotted in Fig. 2.3. The left plot shows the phase $\phi_\tau(L)$ acquired by the fundamental wave at the output of the medium as a function of the normalized input intensity $a_0^2(0)\,\sigma^2$ (in units of $\text{W}\cdot\text{m}^{-2}\ \text{W}^{-1}\cdot\text{m}^2$) for different phase matching conditions. For $\Delta kL = 2\pi$, the highest slope, and therefore highest phase transfer coefficient τ, is obtained at low normalized intensities (smaller than 1.3), where the relationship between phase and normalized input intensity is still linear. At higher intensities, the phase acquired by the fundamental wave inside the medium acts to detune the phase matching condition, which reduces the efficiency of the up and down-conversion steps. As a consequence, the phase ϕ_τ accumulated by the fundamental wave is also

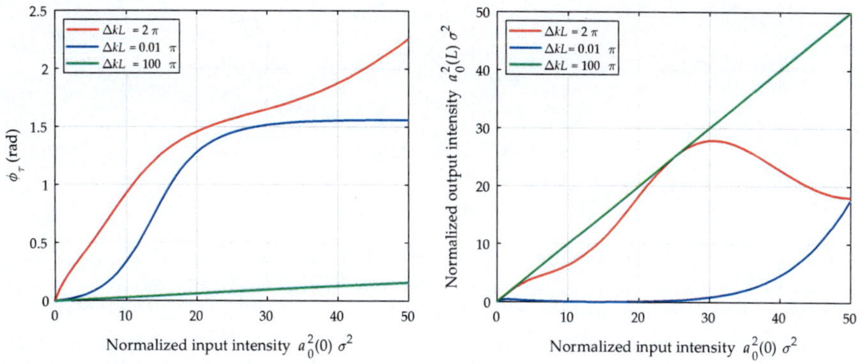

Fig. 2.3 Numerical simulations for the cascaded Kerr effect type I. Left: phase acquired by the fundamental wave at the output of the medium as a function of its normalized input intensity, for three phase matching conditions. Right: normalized output intensity for the fundamental wave as a function of its normalized input intensity

reduced. In this regime, the coherence length will depend on the input intensity of the fundamental wave. The right plot in Fig. 2.3 shows the fundamental normalized intensity at the output of the medium as a function of the input intensity. The relation is linear for low input intensities or for a high phase mismatch ΔkL. Since the phase transfer coefficient and the power in the fundamental beam at the output of the medium are reduced, not much advantage is expected for this transfer scheme in the high intensity regime.

2.1.2.2 Cascaded Kerr Effect Type II

Let us now consider the condition when the input polarization $\hat{e}_0$ of the fundamental wave is neither aligned to the ordinary nor to the extraordinary axis of the nonlinear medium, as shown in Fig. 2.4. From the medium's coordinate system, two fundamental waves, one with extraordinary ($\hat{e}_{0,e}$) and other with ordinary ($\hat{e}_{0,o}$) polarization, are summed and up-converted to a second harmonic wave with ordinary polarization. This condition is known as type II second harmonic generation [47].

Let us assume that the input amplitudes in the ordinary and extraordinary polarization are not equal. In this case, the wave with smaller amplitude, called weak wave, can be completely depleted, while the wave with larger amplitude, called strong wave, will never suffer complete depletion. This is required by energy conservation, since equal energies from the extraordinary and ordinary waves are extracted in the up-conversion step. When the weak wave is completely depleted, the direction of the process will be reversed to down-conversion and the second harmonic and strong waves will restore the weak wave. This reversion forces the weak wave to change its phase during the restoring process, even at a perfect phase matching condition [65, 66]. By controlling the angle of the input polarization, the relative amplitude between the fundamental waves is independently controlled, which is an additional degree

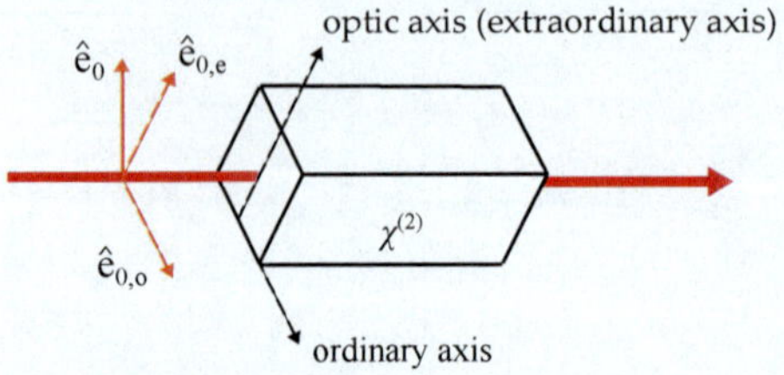

Fig. 2.4 Illustration of the cascaded Kerr effect type II. A fundamental wave propagates through a nonlinear medium whose optic axis is misaligned to the fundamental input field polarization $\hat{e}_0$. From the medium's coordinate system, the polarization can be decomposed into an extraordinary polarization $\hat{e}_{0,e}$, aligned with the optic axis, and an ordinary polarization $\hat{e}_{0,o}$, perpendicular to the optic axis. Each polarization component acquires a different phase shift while propagating through the medium

of freedom for the cascaded process. If the input amplitudes are equal, i.e., input polarization forms an angle of 45° with the optic axis, then the cascaded process is similar to type I and a phase shift is obtained at a phase mismatch condition.

Similarly as in the cascaded type I, the coupled equations for the complex fields are:

$$\begin{aligned}
\frac{\partial \tilde{A}_{0,e}}{\partial \varsigma} &= -i\sigma\, \tilde{A}^*_{0,o}\tilde{A}_2\, e^{-i\Delta k L\varsigma}\,, \\
\frac{\partial \tilde{A}_{0,o}}{\partial \varsigma} &= -i\sigma\, \tilde{A}^*_{0,e}\tilde{A}_2\, e^{-i\Delta k L\varsigma}\,, \\
\frac{\partial \tilde{A}_{2}}{\partial \varsigma} &= -2i\sigma\, \tilde{A}_{0,e}\tilde{A}_{0,o}\, e^{i\Delta k L\varsigma}\,,
\end{aligned} \tag{2.18}$$

with $\Delta k = k_2 - k_{0,e} - k_{0,o}$. This leads to the following equations for the amplitude and phase:

$$\begin{aligned}
\frac{\partial a_{0,e}}{\partial \varsigma} &= \sigma\, a_{0,o}a_2 \sin(\Phi - \Delta k L\varsigma)\,, \\
\frac{\partial a_{0,o}}{\partial \varsigma} &= \sigma\, a_{0,e}a_2 \sin(\Phi - \Delta k L\varsigma)\,, \\
\frac{\partial a_{2}}{\partial \varsigma} &= -2\sigma\, a_{0,e}a_{0,o} \sin(\Phi - \Delta k L\varsigma)\,, \\
\frac{\partial \phi_{0,e}}{\partial \varsigma} &= -\sigma \frac{a_{0,o}a_2}{a_{0,e}} \cos(\Phi - \Delta k L\varsigma)\,, \\
\frac{\partial \phi_{0,o}}{\partial \varsigma} &= -\sigma \frac{a_{0,e}a_2}{a_{0,o}} \cos(\Phi - \Delta k L\varsigma)\,, \\
\frac{\partial \phi_{2}}{\partial \varsigma} &= -2\sigma \frac{a_{0,e}a_{0,o}}{a_2} \cos(\Phi - \Delta k L\varsigma)\,,
\end{aligned} \tag{2.19}$$

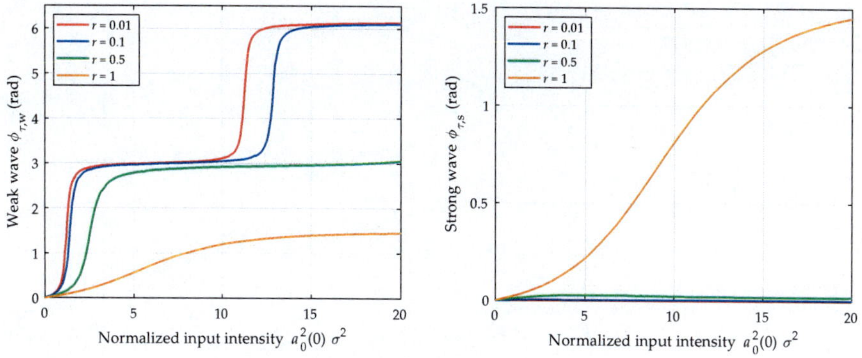

Fig. 2.5 Numerical simulations for the cascaded Kerr effect type II. The phase acquired by the fundamental weak (left plot) and strong (right plot) waves at the output of the medium is plotted as a function of the normalized input intensity of the total fundamental wave, for different input ratios r. In all curves the phase mismatch was set to $\Delta kL = 0.3$ rad

with $\Phi(\varsigma) = \phi_2(\varsigma) - \phi_{0,e}(\varsigma) - \phi_{0,o}(\varsigma)$. The cascading steps for the type II are: $\omega_{0,e} + \omega_{0,o} = 2\omega_{0,e}$, $2\omega_{0,e} - \omega_{0,o} = \omega_{0,e}$, $2\omega_{0,e} - \omega_{0,e} = \omega_{0,o}$.

Figure 2.5 shows the numerical calculation of the phase $\phi_{\tau,w}$ and $\phi_{\tau,s}$ acquired by the weak (here set to be at the ordinary axis) and strong waves at the output of the medium for different ratios $r = a_{0,w}^2(0)/a_{0,s}^2(0) \leq 1$ between the weak and strong input powers. The phase acquired by the weak wave as a function of the normalized input intensity has a step-like behavior, with a period that increases with intensity. The steep slope of the step happens at input intensities in which the regime of regeneration of the weak wave power is initiated, as shown in Fig. 2.6. At those locations, the output intensity of the weak wave is at its minimum value, which is non-zero for $\Delta kL \neq 0$. This is an advantage since only a small fraction of the power in the fundamental wave needs to be used for the sensing process. The steep slope is maximized for small ratios r. For r values below 0.1 no significant difference in the slope is observed, only in the location of the steps which are shifted towards lower normalized input intensities. In addition to that, the power on the weak wave at the locations of the steep slope is considerably reduced for $r < 0.1$. Furthermore, when the input power ratio between the waves is equal, i.e., $r = 1$, both waves acquires identical phase at the output of the medium, and the cascaded effect is similar to the type I condition. The behavior of the phase $\phi_{\tau,w}$ as a function of different phase matching conditions is shown in Fig. 2.7. The acquired phase in each step is always less than π and is reduced with the growth of the mismatch, since the period of energy exchange between the waves is also reduced. In the case of a perfect phase matching condition ($\Delta kL = 0$), not shown in the plot, the weak wave is completely depleted in the up-conversion process and it suffers a discontinuous phase change of π when the process is reversed to down-conversion.

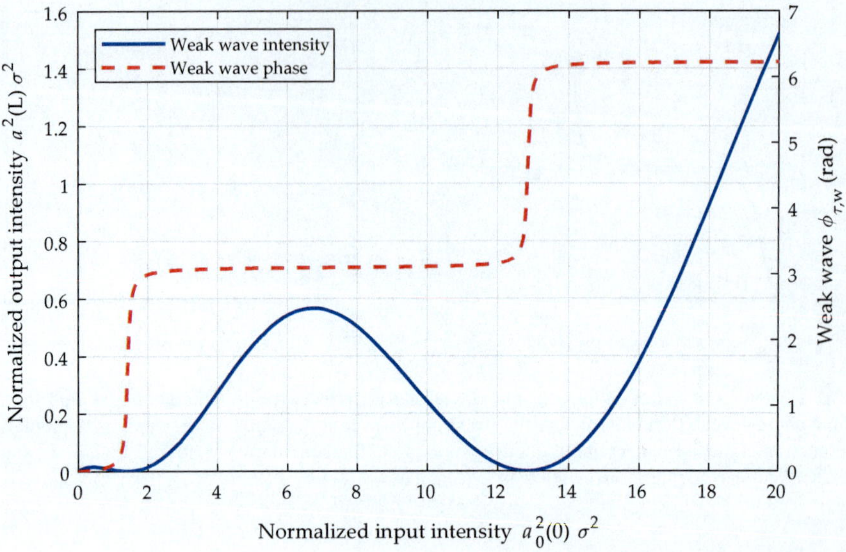

Fig. 2.6 Numerical simulations for the cascaded Kerr effect type II. The normalized output intensity of the weak wave (blue curve) is plotted as a function of the normalized input intensity of the total fundamental wave. The red dashed curve shows, as a reference, the intensity dependent phase acquired by the fundamental weak wave at the output of the medium. In all curves the phase mismatch was set to $\Delta kL = 0.3$ rad and the input power ratio was set to $r = 0.1$

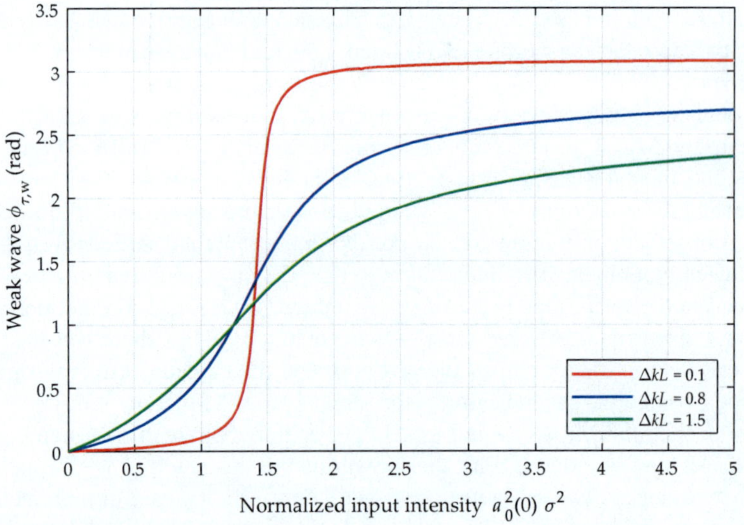

Fig. 2.7 Numerical simulations for the cascaded Kerr effect type II. The phase acquired by the fundamental weak wave at the output of the medium is plotted as a function of the normalized input intensity, for different phase matching conditions ΔkL. For all curves the input power ratio was set to $r = 0.1$

Table 2.3 Phase transfer coefficients τ for the weak wave in the cascading Kerr effect type II transfer scheme, calculated using Eqs. 2.21 and 2.10. The values were calculated considering a beam radius of $w_0 = 20\,\mu\text{m}$, $\lambda_0 = 1064\,\text{nm}$ for KTP, and $\lambda_0 = 1540\,\text{nm}$ for DAST

ΔkL	r	$\frac{\partial\phi_{\tau,s}}{\partial(a_1^2(0)\sigma^2)}$	d_{eff} (pm $\cdot$ V^{-1})	L (mm)	n_2^{eff} (cm^2 $\cdot$ W^{-1})	$\overline{P}_0$ (W)	τ (rad $\cdot$ W^{-1})
0.1	0.1	13.0 at 1.4	3 (KTP [48])	70	7×10^{-15}	7.4	2.4
0.1	0.1	13.0 at 1.4	290 (DAST [61])	1	7×10^{-13}	8.1	2.2
0.1	0.5	5.3 at 2.5	3 (KTP [48])	70	3×10^{-15}	13.2	1
0.1	0.9	1.1 at 6.2	3 (KTP [48])	70	6×10^{-16}	32.8	0.2
0.3	0.1	1.7 at 1.7	3 (KTP [48])	70	9×10^{-16}	9.0	0.3

For small normalized intensity modulations, the effective refractive index and the phase transfer coefficient at a certain input intensity $a_0^2(0)\,\sigma^2$ can be written as:

$$n_2^{\text{eff}} = \frac{\partial\phi_{\tau,\text{w}}}{\partial(a_0^2(0)\sigma^2)} \cdot \frac{4\pi d_{\text{eff}}^2 L}{\lambda_0\epsilon_0 c n_{2\omega_0} n_{\omega_0}^2}, \tag{2.20}$$

$$\tau = \frac{\partial\phi_{\tau,\text{w}}}{\partial(a_0^2(0)\sigma^2)} \cdot \frac{8\pi^2 d_{\text{eff}}^2 L^2}{A\lambda_0^2\epsilon_0 c n_{2\omega_0} n_{\omega_0}^2}. \tag{2.21}$$

Table 2.3 shows the phase transfer coefficient τ for the weak wave as a function of r, ΔkL, type of nonlinear material, and input mean power $\overline{P}_0$. The value of n_2^{eff} was calculated using the literature values of d_{eff} of the named materials. The highest transfer coefficient of 2.4 rad $\cdot$ W^{-1} is obtained with the KTP crystal. This value represents an improvement by a factor of 80 when compared with the cascaded type I condition. The improvement is achieved at the cost of having to operate the experiment at a fixed mean input power of 7.4 W. For these conditions, the power of the weak wave at the output of the medium is 1.1 mW. The next input power leading to a similar slope is around 70 W, and the output power of the weak wave is approximately 1.1 mW (same as for an input power of 7.4 W).

In conclusion, even with a significant improvement with respect to the optical Kerr effect, the phase transfer schemes via cascaded Kerr effect resulted in phase transfer coefficients on the order of 1 rad $\cdot$ W^{-1} or lower. In this case, to detect an RPM of 10^{-9} at frequencies around 10 Hz a power of at least 1 kW is needed. With such a high value in the continuous wave regime, the power absorption in the material cannot be neglected. As a consequence, the phase transfer is reduced, and the damage threshold of most materials will be exceeded for the chosen beam radius of 20 μm. Therefore, longer interaction lengths and higher nonlinearities are required. As mentioned in Sect. 2.1.2.1, high nonlinearities can be achieved with organic materials, such as DAST, which is phase matchable for type I and type II conditions, and N-(4-nitrophenyl)-L-prolinol (NPP) [67, 68]. Those materials are often used for terahertz generation and mode-locking. However, in addition to the limited transparency range for the second harmonic frequency, organic materials are thermally unstable and exhibit lower laser damage threshold in comparison with inorganic materials. Another drawback observed experimentally is the walk-off effect, which reduces the spatial overlap between the fundamental and second harmonic beams as they propagate through the medium, and limits the interaction length L. The combination of these effects compensates for the advantage of an increased nonlinearity and therefore not much improvement is obtained with organic materials. Another important fact to consider is that the analysis was performed considering plane waves. In reality, lasers exhibit a Gaussian beam profile. Since the phase transfer coefficient is inversely proportional to the cross sectional area of the beam, lower coefficients are expected for beams with a Gaussian profile.

In conclusion, the analyzed phase transfer schemes based on the Kerr effect in nonlinear materials are not adequate for a power stabilization in the 10^{-9} stability range and an alternative scheme is required.

2.1.3 Radiation Pressure

The last phase transfer scheme analyzed in this thesis exploits the radiation pressure effect of light on a suspended mirror, as shown in Fig. 2.8. A laser beam impinges on the suspended mirror, and its power modulations δP_0 induce a longitudinal displacement modulation δx in the mirror position via radiation pressure. The mirror displacement modulates the phase of its reflected beam, and therefore a power dependent phase $\delta\phi_\tau$ is obtained.

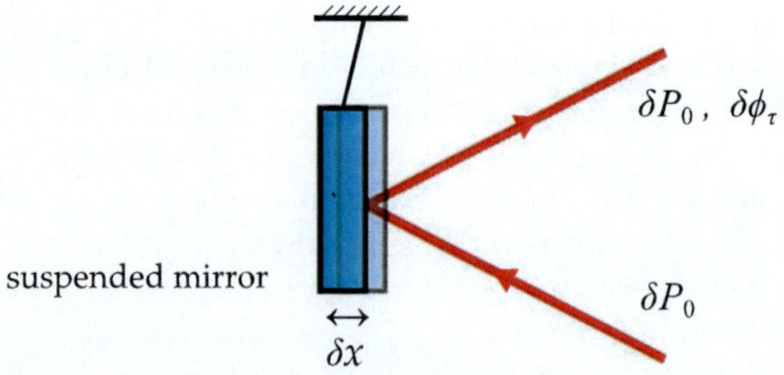

Fig. 2.8 Illustration of the phase transfer scheme via radiation pressure on a suspended mirror. The laser power modulations δP_0 induces modulations δx in the mirror position via radiation pressure. As a result, the phase of its reflected light acquires a power dependent phase modulation $\delta\phi_\tau$

Let us assume that the suspended mirror responds to the radiation pressure force F_{rp} as a damped harmonic oscillator which has the following equation of motion:

$$m\ddot{x}(t) + m\Omega_0^2 (1 + i\phi) x(t) = F_{\mathrm{rp}}(t) , \tag{2.22}$$

where Ω_0 is the mirror fundamental resonance frequency, and m is the fundamental effective mass of the mirror. The complex term $i\phi$ represents the system dissipation, with ϕ being the phase angle in radians by which the response of the oscillator lags behind a sinusoidal driving force, and is therefore called loss angle.

The radiation pressure force is obtained by multiplying the transferred momentum of a single photon upon reflection by the number of photons per second in the laser beam:

$$F_{\mathrm{rp}}(t) = 2 \cdot \frac{\hbar\omega_0}{c} \cdot \frac{P_0(t)}{\hbar\omega_0} = \frac{2P_0(t)}{c} . \tag{2.23}$$

Assuming a linear response to a sinusoidal external force, the equation of motion in the frequency domain is obtained by applying the following Fourier transforms: $x(t) = \int_{-\infty}^{\infty} x(\Omega)e^{i\Omega t}\mathrm{d}\Omega$, and $P(t) = \int_{-\infty}^{\infty} P(\Omega)e^{i\Omega t}\mathrm{d}\Omega$. Then, the mirror position as a function of power is:

$$x(\Omega) = \frac{2P_0(\Omega)}{mc\left(-\Omega^2 + \Omega_0^2 (1 + i\phi)\right)} , \tag{2.24}$$

which has the following magnitude:

$$|x(\Omega)| = \frac{2P_0(\Omega)}{mc\sqrt{\left(\Omega_0^2 - \Omega^2\right)^2 + \Omega_0^4\phi^2}} . \tag{2.25}$$

The radiation pressure transfer function R, defined by:

$$R(\Omega) = \frac{x(\Omega)}{P_0(\Omega)} , \tag{2.26}$$

is plotted in Fig. 2.9 for different masses and loss angles. The magnitude of the transfer function is approximately constant in the low frequency regime:

$$|R(\Omega \ll \Omega_0)| = \frac{2}{cm\Omega_0^2} , \tag{2.27}$$

it reaches its maximum at the resonance frequency, and subsequently falls like $1/\Omega^2$ for high frequencies:

$$|R(\Omega \gg \Omega_0)| = \frac{2}{cm\Omega^2} . \tag{2.28}$$

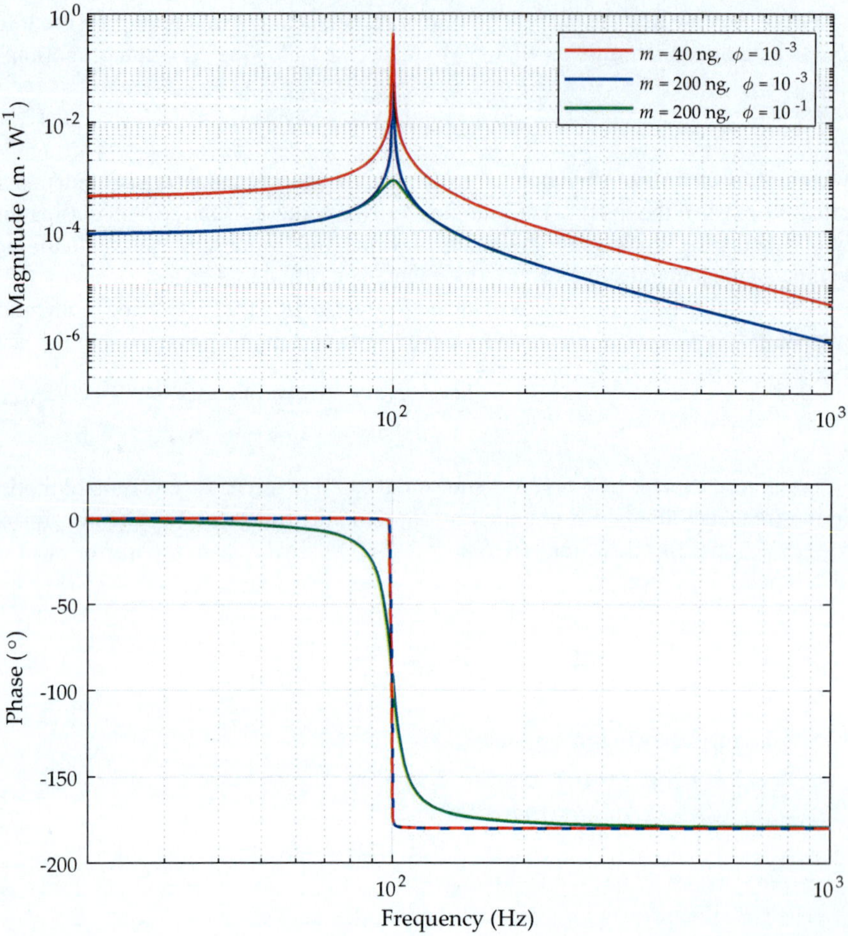

Fig. 2.9 Bode plot of the magnitude and the phase of the radiation pressure transfer function R, calculated from Eq. 2.26, for a suspended mirror with different masses m and loss angles ϕ (here treated as frequency independent). The longitudinal resonance frequency for all plots was set to $\Omega_0 = 2\pi \cdot 100\,\mathrm{Hz}$

The phase of the transfer function is frequency independent and approximately 0° for low frequencies and after resonance it suffers a loss of 180°, which is expected since the oscillator displacement (magnitude of the transfer function) will now decrease with frequency proportionally to $1/\Omega^2$.

The phase ϕ_τ acquired by a reflected beam from the suspended mirror is proportional to twice the mirror displacement and given by $\phi_\tau = 4\pi\,|x|\,/\lambda_0$. The phase transfer coefficient τ of the scheme is therefore frequency dependent and it has the highest value at the resonance frequency. For frequencies below resonance, the coefficient is approximately frequency independent and given by:

Table 2.4 Phase transfer coefficients τ for the radiation pressure transfer scheme for different mirror masses m and longitudinal angular resonance frequencies Ω_0, calculated from Eq. 2.29. A laser with wavelength λ_0 of 1064 nm was considered

References	m (ng)	$\Omega_0/2\pi$ (Hz)	τ $\left(\text{rad} \cdot \text{W}^{-1}\right)$
Corbitt et al. [69]	10^9	10	0.02
Singh et al. [70]	400	274	66
Kleckner et al. [71]	100	10^4	0.2
Thompson et al. [72]	40	134×10^3	3×10^{-3}
Cripe [73]	40	117	3.6×10^3
Verhagen et al. [74]	1.9	7.8×10^7	10^{-7}

$$\tau = \frac{8\pi}{\lambda_0 c\, m\Omega_0^2} . \tag{2.29}$$

The coefficient at frequencies higher than the resonance can be approximated by multiplying Eq. 2.29 with the factor $(\Omega_0/\Omega)^2$.

Table 2.4 shows the phase transfer coefficients for the radiation pressure transfer scheme, calculated from Eq. 2.29, for different parameters found in the literature. The highest coefficient of 3.6×10^3 rad $\cdot$ W^{-1} can be obtained with a micro-oscillator mirror consisting of a 25 μm radius mirror pad suspended by a thin cantilever. This value is 3 orders of magnitude higher than the phase transfer coefficients calculated for the Kerr and cascaded Kerr transfer schemes. At a frequency $\Omega = 1$ kHz, the coefficient is reduced to a value of 50 rad $\cdot$ W^{-1} which is still higher than the coefficients obtained with the previous transfer schemes. For this reason, the radiation pressure is the most promising phase transfer scheme for frequencies up to several kHz with current technologies.

2.2 Phase Readout Schemes

2.2.1 *Michelson Interferometer*

One possibility to measure the phase ϕ_τ is via a Michelson interferometer (introduced in Sect. 2.2.1), as shown in Fig. 2.10. The scheme is illustrated for a self-modulation readout configuration, but it can similarly be implemented for a cross modulation readout. The field transmitted by the beamsplitter acquires a round trip and power dependent phase ϕ_τ, which is measured at the readout port of the interferometer. The output signal of the photodetector is the sensor signal for two control loops: the power stabilization loop, and the length stabilization loop, which keeps the interferometer locked to a specific operational point. An optical isolator, placed at the interferometer

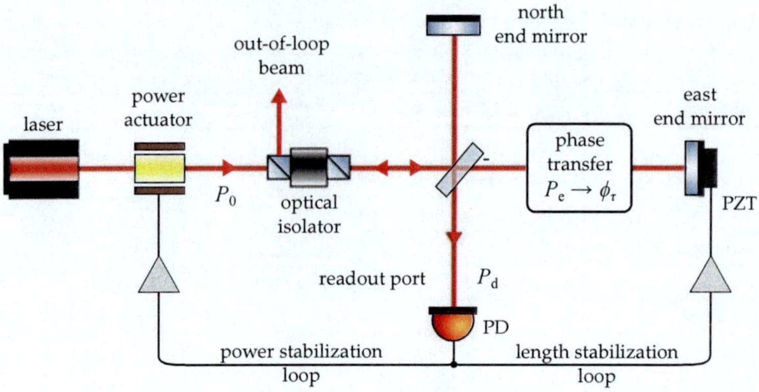

Fig. 2.10 Schematic of laser power stabilization via a phase transfer scheme and a Michelson interferometer readout scheme. The light in the east arm of the interferometer acquires a power dependent phase ϕ_τ which, after interference with light from the north arm, determines the power P_d at the readout port of the interferometer. The photodetector is an in-loop sensor for the power and the length stabilization control loops. An optical isolator reflects the back-reflected beam from the interferometer to the out-of-loop beam

input, reflects the back-reflected beam from the interferometer to the out-of-loop beam, which is stabilized and available for further use.

Let us now calculate the power transfer coefficients η for this scheme. The complex electric field $\tilde{E}_\mathrm{d}$ at the readout port of the interferometer as a function of the round trip phase ϕ_e and ϕ_n, accumulated in the east and north arms, is given by:

$$\tilde{E}_\mathrm{d}\,(t) = \sqrt{R\,T}\,\tilde{E}_0(t)\left(e^{i\phi_\mathrm{n}} - e^{i\phi_\mathrm{e}(t)}\right), \tag{2.30}$$

where $\tilde{E}_0$ is the complex field at the input of the interferometer and R (T) is the reflection (transmission) power coefficient of the beamsplitter. An asymmetric description for the beamsplitter was used, with the reflectivity sign convention shown in Fig. 2.10. Here it was considered that the interferometer end mirrors are lossless and have 100 % reflectivity. This equation can be re-written as a function of the common phase $\overline{\phi}(t) = (\phi_\mathrm{n} + \phi_\mathrm{e}(t))/2$, and the differential phase $\Delta\phi(t) = (\phi_\mathrm{n} - \phi_\mathrm{e}(t))/2$, in the following way:

$$\tilde{E}_\mathrm{d}\,(t) = \sqrt{R\,T}\,\tilde{E}_0(t)\,e^{i\overline{\phi}(t)}\left(e^{i\Delta\phi(t)} - e^{-i\Delta\phi(t)}\right). \tag{2.31}$$

Then, the power P_d at the readout port is:

$$P_\mathrm{d}\,(t) = 4P_0\,(t)\,T\,(1-T)\,\sin^2\left(\Delta\phi\,(t)\right), \tag{2.32}$$

where a lossless beamsplitter was considered, i.e., $R + T = 1$. Because P_d is a function only of the differential arm phase $\Delta\phi$, an interferometer with equal macroscopic

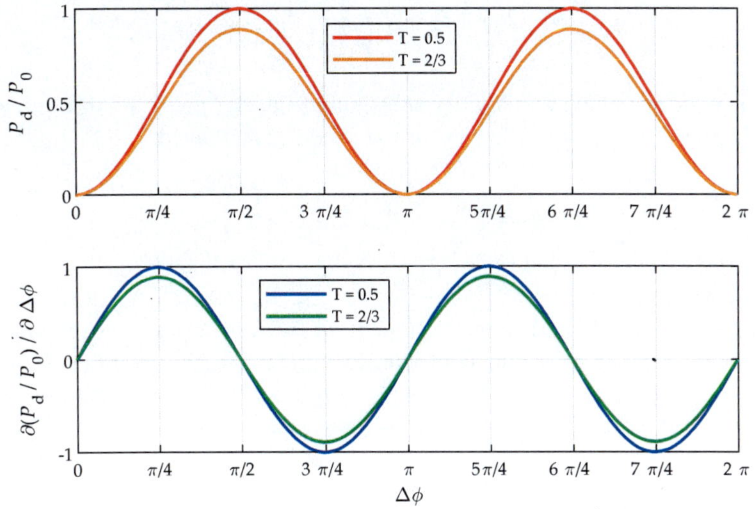

Fig. 2.11 Power response of a Michelson interferometer. Upper plot: power P_d at the readout port of the interferometer, normalized by the input power P_0, as a function of the differential arm phase $\Delta\phi$. Lower plot: derivative of the normalized power P_d/P_0 with respect to the differential arm phase

arm lengths will ideally not couple technical phase noise from the laser source into power noise at its output port, since the phase noise will be common in both interferometer arms. This is an important advantage of performing the phase readout via a Michelson interferometer, since the phase noise of lasers is considerably high at low frequencies (10^2 rad · Hz$^{-1/2}$ at 10 Hz [5]).

Disregarding any noise sources in the interferometer, the differential arm phase can be written as:

$$\Delta\phi\,(t) = \Delta\phi_0 + \delta\phi_\tau\,(t)\ , \tag{2.33}$$

where $\Delta\phi_0$ is the constant phase difference between the interferometer arms, which includes the phase $\phi_\tau\,(\overline{P}_e)$ induced by the mean laser power in the transfer scheme.

The upper plot of Fig. 2.11, shows the power at the interferometer readout port normalized by the input power P_0 as a function of the differential arm phase $\Delta\phi$, for different beamsplitter transmission coefficients. The relation between phase $\delta\phi_\tau$ and power δP_τ modulations is approximately linear for small phase modulations, with a slope that depends on the operational point of the interferometer, i.e., on the constant phase difference $\Delta\phi_0$. The interferometer is locked to a specific operational point by the length stabilization loop, which uses a piezoelectric transducer (PZT) element to actuate in the position of one of the end mirrors.

The highest slope is obtained by locking the interferometer to the mid-fringe, when $\Delta\phi_0$ is an odd multiple of $\pi/4$. Considering small phase modulations, the power $P_{d,mf}$ at the interferometer readout port can be approximated to:

$$P_{\mathrm{d,mf}}(t) \approx 4P_0(t)\,T\,(1-T)\left(\frac{1}{2}+\delta\phi_\tau(t)\right), \tag{2.34}$$

which is a linear function of $\delta\phi_\tau$. By substituting Eqs. 1.3 and 2.3 into Eq. 2.34 the following power transfer coefficient for this scheme is obtained:

$$\eta = 4T\,(1-T)\left(\frac{1}{2}+\overline{P}_0\,T\cdot\tau\right), \tag{2.35}$$

with a mean detected power of $\overline{P}_{\mathrm{d,mf}} = 2T\,(1-T)\,\overline{P}_0$. As expected from operating the interferometer at the mid-fringe, the total power sensing scheme can be interpreted as a sum of a traditional sensing, with a beamsplitter reflectivity of $2T\,(1-T)$ (first term of η), plus a gain provided by the phase transfer scheme (second term of η). As discussed in Sect. 2.1.3, the coefficient τ can reach values of $10^3\ \mathrm{rad}\cdot\mathrm{W}^{-1}$, and therefore $\overline{P}_0\,T\cdot\tau$ can be much larger than 1/2. In this case, η is maximized for $T = 2/3$, which leads to:

$$\eta = 0.6\,\overline{P}_0\cdot\tau\,, \tag{2.36}$$

$$\overline{P}_{\mathrm{d,mf}} = 0.4\cdot\overline{P}_0\,. \tag{2.37}$$

Even though $\eta \gg 1$, a fraction of 0.4 from the laser mean power needs to be detected, which is not desired.

To minimize the mean power at the photodetector, the interferometer can be locked to the dark fringe operational point, by setting $\Delta\phi_0$ to be a multiple of π. In this case, all the laser mean power will be reflected to the out-of-loop beam, and ideally no power is wasted during the sensing process. The dark fringe however, has the disadvantage of having a small response in power. A method to overcome this issue is to perform the readout of the derivative of the interferometer response, which is maximum at the dark fringe, as shown in the bottom plot of Fig. 2.11. The derivative signal can be obtained, for example, by modulating the position of one of the interferometer's end mirror with a sinusoidal wave with amplitude ϵ_{m} and frequency Ω_{m}, which is provided by a local oscillator. For small amplitude modulations, the power at the readout port can be approximated to:

$$P_{\mathrm{d}}\,(\Delta\phi + \epsilon_{\mathrm{m}}\sin(\Omega_{\mathrm{m}}t)) \approx P_{\mathrm{d}}\,(\Delta\phi) + \frac{\partial P_{\mathrm{d}}\,(\Delta\phi)}{\partial\Delta\phi}\cdot\epsilon_{\mathrm{m}}\sin(\Omega_{\mathrm{m}}t)\,. \tag{2.38}$$

The readout is performed from the demodulated signal, which is obtained by first multiplying the photodetector signal by the local oscillator signal, and subsequently by applying a low pass filter, with a corner frequency smaller than Ω_{m}. The resulting signal p_{d} is proportional to:

$$p_{\mathrm{d}} \propto 4T(1-T)P_0\epsilon_{\mathrm{m}}\cos\Delta\phi\cdot\sin\Delta\phi\,, \tag{2.39}$$

which is valid for frequencies $\Omega \ll \Omega_{\mathrm{m}}$. The linear response at the dark-fringe can be obtained by performing a Taylor expansion around $\Delta\phi_0 = \pi$:

$$p_{\mathrm{d,df}} \approx 4T(1-T)P_0\epsilon_{\mathrm{m}}\epsilon_{\mathrm{LO}}\,\delta\phi_\tau\,. \tag{2.40}$$

Hence, the power transfer coefficient of this scheme is:

$$\eta = \epsilon_{\mathrm{m}}\,4T^2(1-T)\overline{P}_0 \cdot \tau\,, \tag{2.41}$$

which is again maximized for $T = 2/3$, and $\overline{P}_{\mathrm{d,df}} \approx 0$.

Readout of the Kerr and Cascaded Kerr Transfer Schemes

It is important to note that for the optical Kerr effect transfer scheme, the factor τ in Eqs. 2.35 and 2.41 needs to be multiplied by a factor of 2 since the beam propagates through the medium 2 times in one arm round trip. This factor should also be included for the cascaded Kerr effect type I transfer scheme, if the scheme is operated in the no-depletion regime and with a phase matching condition of $\Delta kL = 2\pi$ (ideally no second harmonic light comes out of the medium). If the scheme is operated in the high intensity regime, the depletion of the fundamental field after the first propagation through the medium must be taken into account in the calculation of τ for the second pass in the medium. Additionally, the power loss for the fundamental field in the interferometer arm also needs to be included in the calculation of η. An example of calculations for the cascaded Kerr effect type I with a double pass configuration and a dichroic mirror can be found in [75]. The readout for the cascaded Kerr effect type II transfer scheme is not optimal with a Michelson interferometer. This is because the power dependent phase of interest is induced in the weak wave, which has a polarization misaligned with respect to the polarization of the reference wave at the north arm. In addition to that, as it will be shown in Sect. 2.3, the output polarization of the fundamental wave after the cascaded scheme will, in general, be elliptical. For this scheme, an optical cavity, discussed in the following section, is a better readout configuration.

Readout of the Radiation Pressure Transfer Scheme

To realize the readout from the radiation pressure transfer scheme, the east end mirror should be substituted by the suspended mirror. In this case, no additional factor has to be taken into account, since τ was already calculated in reflection of the suspended mirror. The readout of the radiation pressure scheme can alternatively be performed for a cross phase modulation, as shown in Fig. 2.12. In this configuration, the power modulations δP_{t} of a strong beam, here called transfer beam, modulates the position of the suspended mirror, which in turn modulates the phase in the field propagating through the interferometer east arm. The beam in the interferometer, called the sensing beam, has a mean power $\overline{P}_{\mathrm{s}}$ much smaller than the power of the transfer beam, and therefore its self phase modulation effect will be, for now, neglected. Considering that the sensing beam has the same wavelength of the transfer beam, the induced phase modulation in the interferometer's east arm field is

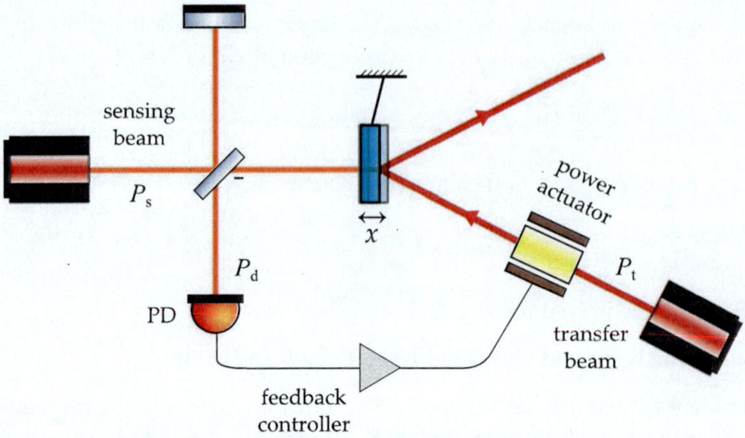

Fig. 2.12 Schematic of laser power stabilization scheme via radiation pressure, with a cross phase modulation configuration. The power modulations δP_t of a strong transfer beam, modulates the position of the suspended mirror, and consequently the phase in the field propagating in the east interferometer arm. The power modulations δP_τ at the readout port of the interferometer are sensed by the photodetector and fed back to a power actuator, which stabilizes the transfer beam power

$\delta\phi_\tau = \tau \cdot \delta P_t$. In this configuration, the power transfer coefficient from power modulations of the transfer beam to power modulations at the readout port of the interferometer, locked to the dark-fringe, is:

$$\eta = \epsilon_m \epsilon_{LO} 4T(1-T)\overline{P}_s \cdot \tau \,, \tag{2.42}$$

which is maximum for $T = 0.5$, and smaller than the coefficient from Eq. 2.41 (self modulation configuration) since $\overline{P}_s \ll \overline{P}_0 = \overline{P}_t$. A calculation of the fundamental limits for this scheme can be found in Chap. 3.

Mach-Zehnder Interferometer

The analysis carried out for the Michelson interferometer can be extended to other interferometer configurations, such as a Mach-Zehnder interferometer. An advantage for the Mach-Zehnder is that both interferometer output ports are free for usage, and no optical isolator is needed to separate the out-of-loop beam. This configuration however, might not be ideal for the radiation pressure transfer scheme since a normal angle of incidence for the suspended mirror is not possible. To my knowledge, the first proposal and experimental demonstration of a nondemolition measurement in combination with a negative feedback to reduce laser power fluctuations was reported in 1986 [76]. In this experiment, the power modulations of a strong transfer beam, to be stabilized, modulated the refractive index of a medium via the Kerr effect, as illustrated in Fig. 2.13. A weak sensing beam, with an optical frequency $\omega' \neq \omega_0$, propagated through the Kerr medium and acquired a phase ϕ_τ, modulated by the power of the strong beam (cross phase modulation). The phase ϕ_τ of the weak beam

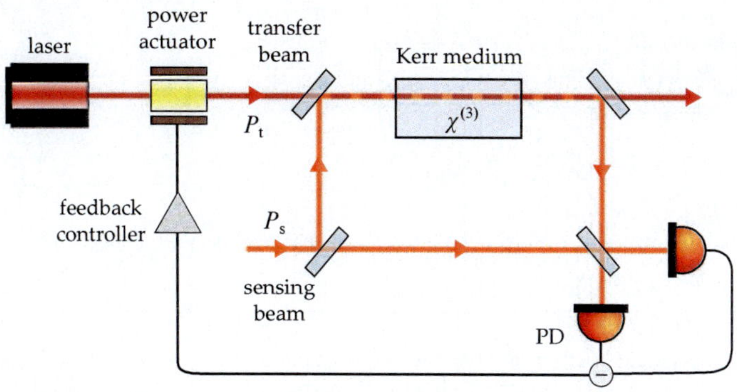

Fig. 2.13 Schematic of laser power stabilization via a optical Kerr transfer scheme and a Mach-Zehnder interferometer readout scheme. The power modulations of a strong transfer beam, modulates the index of refraction of a Kerr medium. The cross phase modulations induced in a weak sensing beam are sensed with a Mach-Zehnder interferometer plus a homodyne detector and fed back to a power actuator, which stabilizes the transfer beam power

was then measured by a Mach-Zehnder interferometer and the output signal was fed back to the transfer beam. As a result, a stabilization between 5 and 10 dB below the relative shot noise of the strong laser was observed in the MHz regime. At lower frequencies, the experiment was limited by technical noise sources. The advantage of measuring the cross phase modulation induced in the sensing beam instead of the self phase modulation of the strong beam, is that the refractive index n_2 in this case is twice as large due to a degeneracy factor in the calculation of the third order susceptibility $\chi^{(3)}(\omega' = \omega' + \omega_0 - \omega_0)$ [47].

2.2.2 *Optical Cavity*

Transfer Scheme Outside the Cavity

Another possibility to readout the phase ϕ_τ is to use an optical cavity, as shown in Fig. 2.14. Here, the beam at the output of the transfer scheme is sent to a linear two-mirror cavity (Fabry-Péot cavity [77, 78]), which has its length controlled by a PZT actuator. The light field will be resonant in the cavity when the cavity length is an integer number of $\lambda_0/2$. Hence, if the cavity length is fixed, variations in the phase of the incoming field can be detected by measuring variations in the resonance condition. There are several techniques to lock the cavity length to a specific operational point and to perform the phase readout. The Pound-Drever-Hall technique [79, 80] for example, performs a null measurement in reflection of the optical cavity, similar to what was described for the dark fringe readout in the Michelson interferometer. A modulator applies a periodic phase modulation in the laser field at a fixed frequency Ω_m, here chosen to be high enough so that the induced phase modulation sidebands

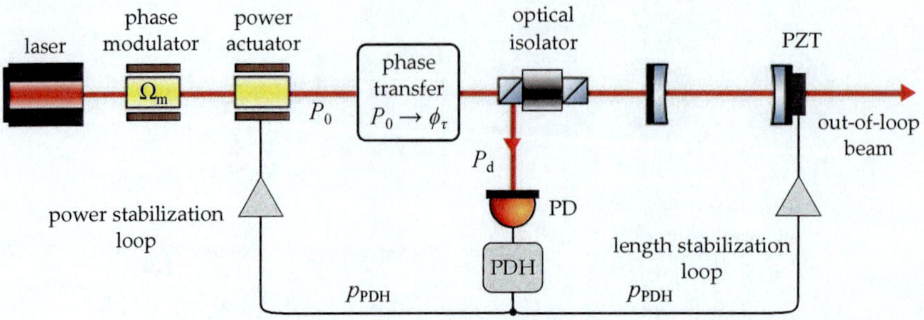

Fig. 2.14 Schematic of laser power stabilization via a phase transfer scheme and an optical cavity readout scheme. The power dependent phase modulations $\delta\phi_\tau$ are measured in reflection of an optical cavity via the Pound-Drever-Hall (PHD) technique. The signal p_{PDH} is used as an error signal for the power and for the length stabilization control loops

are totally reflected by the cavity, regardless on the cavity resonance condition. A photodetector, placed in reflection of the cavity, measures the optical beat between the reflected carrier field, which depends on the resonance condition, and the fixed modulation sidebands. On resonance, the beat signals are 180° degrees out of phase and are canceled, while off resonance they produce a power modulation at a frequency Ω_m. A demodulation technique, similar to what described for the Michelson interferometer dark-fringe locking, is used to generate a signal p_{PDH}, which is the error signal for the power stabilization and for the length stabilization control loops.

Let us consider a high finesse cavity which is locked near resonance. Then, for phase modulations $\delta\phi_\tau$ at frequencies Ω much smaller than the cavity linewidth $\delta\nu$, the error signal p_{PDH} is proportional to [80]:

$$p_{PDH} \propto \frac{4\Omega\,(P_c P_s)^{1/2}}{\pi\,\delta\nu} \cdot \delta\phi_\tau\,, \tag{2.43}$$

where P_c is the power in the carrier, and P_s is the power in each modulation sideband. Here it was assumed that the modulation amplitude is small such that the approximation $P_c + 2P_s \approx \overline{P}_0$ is valid. To derive this equation, the following relationship between modulations $\delta\omega$ in the optical frequency and in the phase was used:

$$\delta\omega = \delta\phi \cdot \Omega\,. \tag{2.44}$$

The power transfer coefficients for this scheme are then proportional to:

$$\eta = \frac{4\Omega\,(P_c P_s)^{1/2}}{\pi\,\delta\nu} \cdot \tau\,, \tag{2.45}$$

and $\overline{P}_d \approx 0$ for an impedance matched cavity. The coefficient η is frequency dependent, and from the assumptions made to derive this equation, $\Omega/\delta\nu \ll 1$. Hence, the power transfer coefficient in this configuration will be smaller than for the Michelson interferometer.

Equation 2.43 was derived assuming that the time dependent phase of the laser field at the input of the cavity has only the contribution from the phase modulations $\delta\phi_\tau$. In reality, the phase noise $\delta\phi_\ell$ from the laser source must also be taken into account, by doing the substitution $\delta\phi_\tau \rightarrow \delta\phi_\tau + \delta\phi_\ell$, since there is no distinction from these terms by the cavity. Hence, laser phase noise will couple at the power readout, which is a disadvantage of using an optical cavity over a Michelson interferometer.

To perform the readout from the cascaded Kerr effect type II, a PBS has to be placed after the transfer scheme in order to split the weak beam, that will be sent to the cavity, from the strong beam, that will be used as an out-of-loop beam. Since the power on the weak beam is much smaller than the input power $\overline{P}_0$, the gain in the factor τ obtained with this transfer scheme will be compensated by the reduction in the readout term $P_c P_s$. This is an important fact to be taken into account in the choice of the transfer and readout schemes, since the figure of merit in the stabilization scheme is the power transfer coefficient η.

Transfer Scheme Inside the Cavity

A benefit of using an optical cavity is to exploit the high intracavity power P_{cav}, by placing the phase transfer scheme inside the cavity. In the steady state regime, the intracavity beam can be described by a standing wave with a power dependent phase $\phi_\tau(t) = \tau \cdot P_{cav}(t)$, which is the same phase obtained as if a traveling wave with power P_{cav} would propagate only once through the transfer scheme. The intracavity U_{cav}, the reflected U_r, and the transmitted U_t fields, considering a lossless cavity, are given by [81]:

$$
\begin{aligned}
U_{cav}(t) &= \frac{i\sqrt{1-R_1}}{1-\sqrt{R_1 R_2}\, e^{2i\left(\Delta\phi_0+\tau|U_{cav}(t)|^2\right)}} \cdot U_0(t)\,, \\
U_r(t) &= \sqrt{R_1} \cdot U_0(t) + i\sqrt{1-R_1} \cdot U_{cav}(t)\,, \\
U_t(t) &= i\sqrt{1-R_2} \cdot U_{cav}(t)\,,
\end{aligned}
\tag{2.46}
$$

where R_1 and R_2 are the power reflectivity coefficients for the input and end mirrors, respectively, and $\Delta\phi_0$ is the static cavity detuning from resonance. For this calculation, a symmetric description for the cavity mirrors were used. The fields here were normalized such that $|U_j|^2$ has units of Watts. Hence, for $\tau \neq 0$, the intracavity power is a nonlinear function of the input power P_0:

$$
P_{cav}(t) = \frac{1-R_1}{1-R_1R_2-2\sqrt{R_1R_2}\cos\left(2\Delta\phi_0+2\tau P_{cav}(t)\right)} \cdot P_0(t)\,. \tag{2.47}
$$

This equation was solved numerically for different cavity detunings, and the result is shown in the left plot of Fig. 2.15. The blue curve was plotted for a cavity without the transfer scheme, which is simulated by simply setting $\tau = 0$ into Eq. 2.47. The

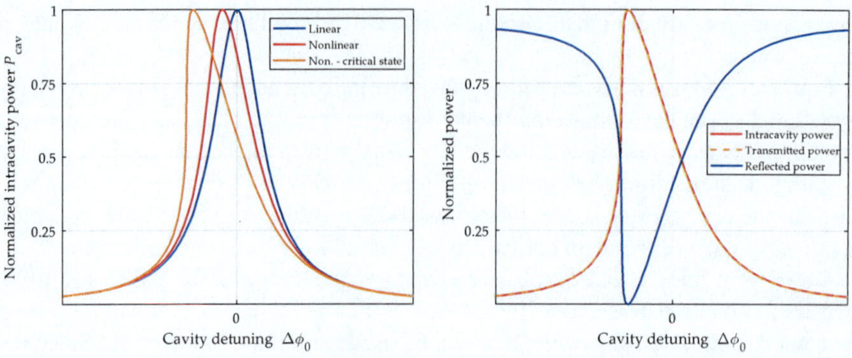

Fig. 2.15 Numerical simulations for a nonlinear optical cavity. Left: intracavity power, normalized to 1, as a function of the cavity detuning. The blue curve corresponds to a linear cavity ($\tau = 0$ in Eq. 2.47), while the red and orange curves correspond to a nonlinear cavity, with the transfer coefficient τ for the yellow curve 3 times higher than for the red curve. Right: normalized powers simulated for a nonlinear cavity in the critical state. The red curve displays the normalized intracavity power, the dashed yellow curve the transmitted power, and the blue curve the reflected power

red and the orange curves were plotted for a nonlinear cavity, with a positive phase shift introduced by the transfer scheme, and with a transfer coefficient τ for the yellow curve 3 times higher than for the red curve. The curves show that, due to the positive additional constant phase $\phi_\tau(\overline{P}_{\text{cav}})$ induced by the mean intracavity power, the resonance condition is achieved at a negative detuning compared to the cavity without the phase transfer scheme ($\tau = 0$). In addition to that, higher slopes between power and phase can be achieved in a nonlinear cavity, with values close to infinity for a detuning corresponding to 0.75 of the maximum intracavity power, in the so called critical state (yellow curve). As a consequence, the power transfer coefficient η is significantly enhanced in this configuration, which can also be seen in the behavior of the reflected power by the cavity shown by the blue curve in the right plot of Fig. 2.15. However, the coupling of phase noise into the readout will be equally enhanced in the critical state, and no advantage is gained by placing the transfer scheme inside the cavity from this point of view.

A passive power stabilization scheme using the cascaded Kerr effect type I transfer scheme and an optical cavity was reported in [82, 83]. The experiment implemented a 6.3 mm long MgO:LiNbO$_3$ crystal ($d_{\text{eff}} \approx 5.3\,\text{pm} \cdot \text{V}^{-1}$) with a transfer factor coefficient τ of 4.5×10^{-3} rad $\cdot$ W^{-1}. The crystal was placed inside the optical cavity, which was operated approximately in the critical state. The passive noise reduction of this scheme is a consequence of the correlation between phase and amplitude fluctuations introduced by the transfer scheme, which transforms an assumed circular quadrature noise distribution of the laser into an ellipse (more details about this approach will be discussed in Sect. 3.5). In reflection of the cavity, the semi-minor axis of this ellipse, which has a smaller value than the noise in the amplitude quadrature of the laser, is rotated such that is aligned to the laser carrier. As a result, a technical power noise reduction of 32 dB at frequencies around 0.9 MHz was observed in reflection of the cavity by a self-homodyne detector. At lower frequencies, the measurement showed a non stationary noise that increased with the cascaded Kerr effect, and was attributed to internal Brillouin scattering [83]. Additionally, the authors also cite the possibility of thermally driven noise, since the intracavity power reached values on the order of 100 W at a beam waist size of 30 μm. The experiment predicted a maximum of 6dB amplitude squeezing, which was not observed due to high amount of classical phase noise of the laser. The configuration of this experiment is not ideal for the purposes of this thesis, since the power noise reduction occurs only in reflection of the cavity, meaning that only 0.25 of the input power has its power stabilized.

2.3 Polarization Transfer and Readout Schemes

This section presents transfer schemes in which the output polarization state of the light is dependent on the input laser power, as illustrated in Fig. 2.16.

A wave, propagating in the z direction, and linearly polarized along the x-axis, is sent to a transfer unit which introduces a power dependent change in the polarization of the output electric field $\overrightarrow{E}_\tau$ in the following way:

$$\overrightarrow{E}_\tau = E_\tau \cos\left(kz + \phi_\tau\right)\hat{e}_\tau = E_{\tau,x}\cos\left(kz + \phi_{\tau,x}\right)\hat{x} + E_{\tau,y}\cos\left(kz + \phi_{\tau,y}\right)\hat{y}\,, \tag{2.48}$$

where $E_{\tau,x}$ ($E_{\tau,y}$) is the power dependent $x(y)$-component of the amplitude at the output of the transfer unit, and $\phi_{\tau,x}$ ($\phi_{\tau,y}$) the corresponding power dependent phase. The output polarization will remain linear if the phase difference $\phi_{\tau,x} - \phi_{\tau,y}$ is equal to zero or an integer multiple of $\pm\pi$. In this case, the polarization state $\hat{e}_\tau$ is rotated with respect to the input polarization by an angle $\theta = \text{atan}\left(E_{\tau,y}/E_{\tau,x}\right)$, as shown in Fig. 2.16a. For other values of the phase difference, the output polarization state will be elliptical and the angle θ will be defined as the angle between the semi-major axis of the ellipse and the input polarization, as shown in Fig. 2.16b, and given by:

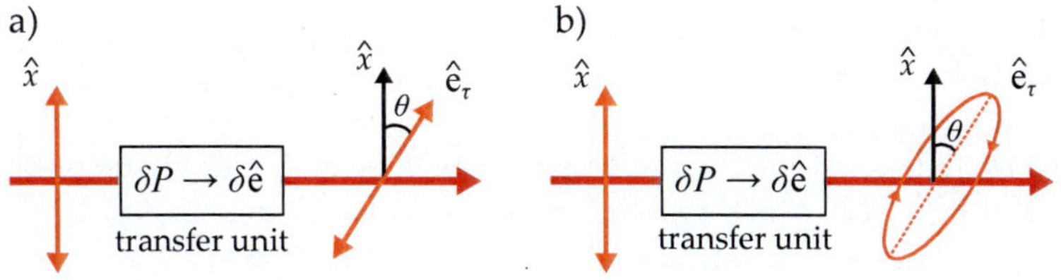

Fig. 2.16 General concept of a polarization transfer unit: an input wave linearly polarized along the x axis experiences a power dependent change in its polarization state after the transfer unit. As a result, the output wave will have a linear (**a**) polarization state which is rotated by an angle θ, or an elliptical (**b**) polarization with a semi-major axis forming an angle θ with the x axis

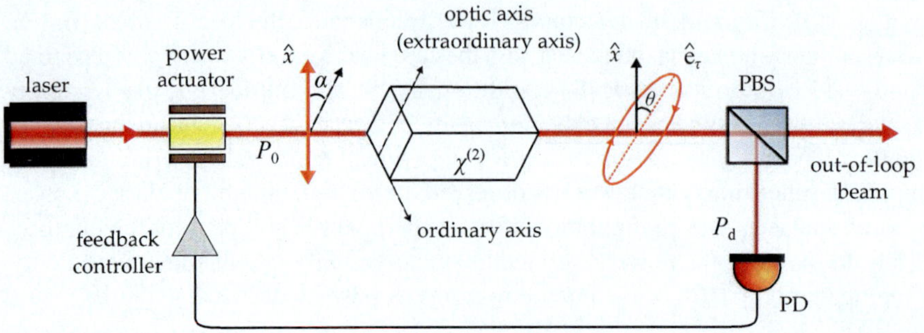

Fig. 2.17 Schematic of laser power stabilization via a polarization transfer scheme exploiting the cascaded Kerr effect. A second order nonlinear medium, which has its optical axis rotated by an angle α with respect to the input polarization, induces a power dependence in the polarization state of the light. The polarization state is detected in reflection of a PBS by a photodetector, whose output signal is fed back to a power actuator

$$\tan 2\theta = \frac{2E_{\tau,x}E_{\tau,y}\cos\left(\phi_{\tau,x} - \phi_{\tau,y}\right)}{E^2_{\tau,x} - E^2_{\tau,y}}. \tag{2.49}$$

One option to induce a power dependence in the polarization state of the wave is via the cascaded Kerr effect, introduced in Sect. 2.1.2. In this case, the transfer occurs via a second order nonlinear and birefringent medium which has its optical axis rotated by an angle α with respect to the input polarization $\hat{x}$, as shown in Fig. 2.17. The input polarization is decomposed into an ordinary and extraordinary component, with corresponding amplitudes and phases that are affected differently inside the medium, as a function of the input power. Hence, a power dependent change in the output polarization is introduced. As a consequence, power modulations δP_0 at the input light will induce modulations $\delta\theta$ that are converted to power modulations δP_τ in reflection of a PBS. The modulations δP_τ are sensed by the in-loop photodetector and fed back to the power actuator via the feedback controller.

If the medium is phase matched for the cascaded Kerr effect type I condition, the field component with polarization aligned to the ordinary (or extraordinary) axis is kept at a phase matching condition and acquires an intensity dependent phase shift. Additionally, this field will also have its amplitude depleted as a function of intensity if the cascaded effect occurs in the high intensity regime. The orthogonal polarization however, is kept far from phase matching and it ideally propagates in the linear regime. As a result, at the output of the medium the two polarizations will combine to a field with an intensity dependent polarization state. If the medium is phase matched for the cascaded Kerr effect type II condition, the fields in the ordinary and extraordinary polarizations will be ruled by the same phase matching condition but will acquire different phases and/or suffer different depletion during propagation. The result is again an intensity dependent polarization at the output of the medium.

Here the polarization will be analyzed only for the type II condition since, from the analysis of Sect. 2.1.2 and from [84, 85], the effect with the type I condition is known to be smaller.

The output fundamental field $\tilde{A}_\tau$, projected to the xy coordinate system, can be obtained from the individual Jones matrices [86] of the scheme:

$$\begin{bmatrix}\tilde{A}_{\tau,x}\\ \tilde{A}_{\tau,y}\end{bmatrix} = \begin{bmatrix}\cos\alpha & -\sin\alpha\\ \sin\alpha & \cos\alpha\end{bmatrix}\begin{bmatrix}\rho_e(L)e^{i\phi_e(L)} & 0\\ 0 & \rho_o(L)e^{i\phi_o(L)}\end{bmatrix}\begin{bmatrix}\cos\alpha & \sin\alpha\\ -\sin\alpha & \cos\alpha\end{bmatrix}\begin{bmatrix}\tilde{A}_{0,x}(0)\\ 0\end{bmatrix}$$
$$= \tilde{A}_{0,x}(0)\begin{bmatrix}\rho_e(L)e^{i\phi_e(L)}\cos^2\alpha + \rho_o(L)e^{i\phi_o(L)}\sin^2\alpha\\ \cos\alpha\sin\alpha\left(\rho_e(L)e^{i\phi_e(L)} - \rho_o(L)e^{i\phi_o(L)}\right)\end{bmatrix},$$

where ρ_o and ρ_e are the transmission coefficients for the amplitude in the ordinary and extraordinary polarizations, respectively. Let us assume that the PBS reflects polarization aligned to the y axis, and transmits polarization aligned to the x axis. Then, if the PBS forms an angle β with respect to the x axis, the reflected field $\tilde{A}_r$ by the beamsplitter will be:

$$\tilde{A}_r = \begin{bmatrix}0 & 0\\ 0 & 1\end{bmatrix}\begin{bmatrix}\cos\beta & \sin\beta\\ -\sin\beta & \cos\beta\end{bmatrix}\begin{bmatrix}\tilde{A}_{\tau,x}\\ \tilde{A}_{\tau,y}\end{bmatrix} \tag{2.50}$$
$$= -\tilde{A}_{\tau,x}\sin\beta + \tilde{A}_{\tau,y}\cos\beta\,. \tag{2.51}$$

For the proposed power sensing scheme, the angle β needs to be chosen as a trade-off between small mean power reflected by the beamsplitter and high power transfer coefficient η for the power modulations. Let us define an additional coefficient to quantify the ratio between the mean main and detected powers, $\eta_{DC} = \overline{P}_d/\overline{P}_0$.

Figure 2.18 shows the normalized output intensity in the x (left plot) and y (right plot) polarizations as a function of the normalized input intensity for different ratios r. The relation used between the angle α and the ratio r between the ordinary (weak) and extraordinary (strong) waves input power was:

$$\alpha = \arctan\left(\sqrt{r}\right)\,. \tag{2.52}$$

When $r < 0.2$, the output wave is kept predominantly in the x polarization. For $r > 0.2$, the output wave is predominantly at the y polarization for low input normalized intensities, and predominantly at the x polarization for high intensities. The behavior of the output intensities depending on the phase mismatch parameter ΔkL is shown in Fig. 2.19. From the curves it is possible to see that the optimum conditions to increase η and decrease η_{DC} depends on the chosen value for the mean normalized input intensity.

Table 2.5 presents the power transfer coefficients η and η_{DC}, calculated for $\beta = 0$. It also shows the output normalized intensity in the x and y polarization, as well as in the second harmonic wave. The best compromise between a high η and a low η_{DC} was obtained for an operational point of $a_0^2(0)\,\sigma^2 = 16$, a perfect phase

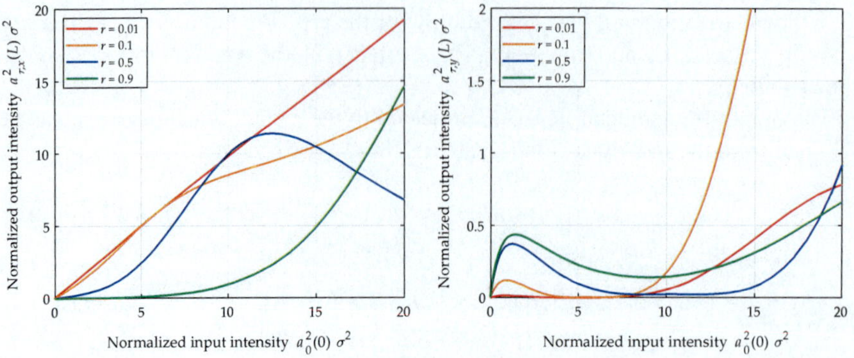

Fig. 2.18 Numerical simulations for the polarization transfer scheme via cascaded Kerr effect type II: normalized output intensity of the fundamental wave in the x (left plot) and y (right plot) polarizations for different input ratios r, and for a phase mismatch of $\Delta kL = 0.1$ rad

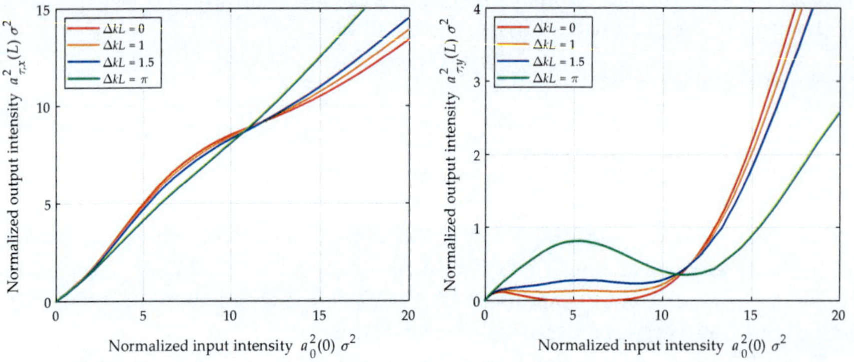

Fig. 2.19 Numerical simulations for the polarization transfer scheme via cascaded Kerr effect type II: normalized output intensity of the fundamental wave in the x (left plot) and y (right plot) polarizations for different phase matching conditions ΔkL, and for a power ratio $r = 0.1$

Table 2.5 Power transfer coefficients η and η_{DC} calculated numerically for the cascaded Kerr effect type II polarization transfer scheme for $\beta = 0$. A beam with radius of $w_0 = 20\,\mu$m, and λ_0 = 1064 nm, was considered

ΔkL (rad)	r	$a_0^2(0)\,\sigma^2$	η	η_{DC}	$a^2_{\tau,x}(L)\,\sigma^2$	$a^2_{\tau,y}(L)\,\sigma^2$	$a_3^2(L)\,\sigma^2$
0	0.1	0.15	0.24	0.26	0.1	0.04	6×10^{-3}
0	0.5	0.15	0.65	0.8	0.01	0.12	0.02
0	0.1	16	0.75	0.17	11	2.8	2.2
1.5	0.1	20	0.62	0.25	14.5	5	0.5
1.5	0.2	22	1.10	0.36	12.2	8	1.8
0.1	0.2	9	0.70 ($2\omega_0$)	0.09 ($2\omega_0$)	8.1	0.1	0.8

matching condition, and a power ratio of 0.1, in accordance with Figs. 2.18 and 2.19. Considering a 70 mm long KTP crystal, with $d_{\text{eff}} = 3\,\text{pm} \cdot \text{V}^{-1}$, and a 1064 nm beam with radius of 20 μm, the power transfer coefficient η of 0.75 is obtained at an input power of 90 W, and results in 15 W of detected power in the reflected y polarization, which is not desired. Higher factors η can be obtained at higher input intensities, at the cost of also increasing the coefficient η_{DC}, and therefore the detected power, as shown in row number five of the table. The same situation would be true in the case where $\beta \neq 0$. The last row of the table also shows an alternative power sensing option in which the power fluctuations of the input beam are inferred by measuring the power fluctuations in the output second harmonic beam. This option has the best compromise between the transfer coefficients η and η_{DC}. If again the same parameters for the KTP crystal are used, then the coefficients are obtained detecting 4.2 W of power in the second harmonic wave, which is still very high.

Experimental demonstrations of intensity dependent polarization states were reported for cascaded Kerr effect type II [85, 87] and type I [85]. The results presented in the literature agreed well with numerical simulations and the small observed discrepancies were attributed to walk-off effects and group velocity dispersion. These experiments were performed with pulsed lasers, which exhibit higher peak powers and experience a higher power damage threshold than continuous wave lasers. For the continuous wave regime, the damage threshold of the materials are smaller and might not be sufficient to reach operational points where η is sufficiently large. The same discussion as presented in the end of Sect. 2.1.2 applies to this scheme regarding how the power transfer coefficients can be increased and regarding the limitations of the nonlinear materials.

The calculations for an intensity dependent polarization state induced via the optical Kerr effect are more complex than for the cascaded Kerr effect, due to the fourth-rank tensor properties of the third order susceptibility. Detailed calculations considering an isotropic medium were carried out in [47]. The calculations show that the polarization state of an input light which is linearly or circularly polarized, is unchanged upon propagation in the nonlinear isotropic medium. Only a wave with input elliptical polarization will suffer a polarization rotation. The rotation angle will be proportional to the difference between the intensities of the left- and right-hand circular components of the input wave and therefore it is not an appropriate measurement to induce the total power fluctuations of the input wave.

One limitation in the sensitivity that can be achieved via a polarization transfer scheme is polarization noise. This coupling can be minimized by placing a polarization filter before the transfer scheme, for example. The polarization noise however cannot be reduced, with classical methods, below the fundamental limit set by the uncertainty in the Stokes parameters [86, 88] of the laser. After the transfer scheme, the polarization noise cannot be filtered independently from the induced polarization modulations, and it will be inevitably converted into power noise in reflection and transmission of the PBS, and set a limit to the power stability that can be achieved in the out-of-loop beam.

It is interesting to note that schemes with an intensity dependent polarization state can be used passively (without feedback) to generate polarization squeezing. For example, Kerr effect in a 13.3 m polarization maintaining fiber was used to generate bright polarization squeezing of up to 5.1 dB [89]. The experiment was perfomed with a Cr^{4+}:YAG laser emitting 130-fs pulses at 1497 nm at a repetition rate of 163 MHz. The beam was coupled into the two orthogonal polarization axes of the fiber. At the output of the fiber, the beams interfered and, as a result, polarization squeezing was identified by measuring the Stokes parameters with a scheme employing a PBS. Other reports on polarization squeezing can be found in the literature, such as squeezing at 1064 nm and in the continuous wave regime [90], squeezing using a KTP crystal phase matched for type II condition [91], and squeezing using photonic crystal fibers [92]. With the feedback scheme proposed in this section, active squeezing in the polarization seems not to be possible, since the feedback control loop acts only in the amplitude of the light.

2.4 Summary

This chapter analyzed an alternative scheme to sense laser power fluctuations. In this scheme, the power fluctuations are transferred to another observable of the light field, which are then detected by a readout scheme.

The following transfer schemes from power to phase were analyzed: optical Kerr effect, cascaded Kerr effect type I and type II, and radiation pressure. These schemes were compared with each other using the phase transfer factor τ as a figure of merit, which expresses the ratio between the induced phase and the power, in units of $\text{rad} \cdot \text{W}^{-1}$. The highest τ, of $3.6 \times 10^3\ \text{rad} \cdot \text{W}^{-1}$, was obtained for the radiation pressure transfer scheme with a micro-oscillator mirror with a mass of 40 ng and a longitudinal resonance frequency of 117 Hz. This coefficient is at least three orders of magnitude higher than the corresponding coefficient achieved with the Kerr and cascaded Kerr transfer schemes. In addition to that, experiments with Kerr and cascaded Kerr effects reported in the literature revealed additional noise originating from nonlinear behavior and from the required for high power, like stimulated Brillouin scattering and thermal fluctuations. Hence, it was concluded that, with current technologies, the signal to noise ratio using Kerr and cascaded Kerr effects would not be sufficient for a proof-of-principle experiment at frequencies below the kHz regime, and that the radiation pressure scheme is the most promising option. An analysis of a polarization transfer scheme based on the cascaded Kerr effect resulted also in a low signal to noise ratio, due to the same mentioned reasons for the phase transfer scheme, and therefore it is also not appropriate for the purposes of this thesis.

The most suitable readout scheme for the phase induced by radiation pressure, for a proof-of-principle experiment, is a Michelson interferometer. The interferometer has the advantage of not being sensitive to laser phase noise, which dispenses the need of a phase stabilization scheme. In addition to that, due to the high transfer coefficient with a micro-oscillator, the experiment will not be limited by power availability which spare the need of an intracavity power build-up. Finally, the interferometer is simpler to operate and to align than a cavity which, with a micro-oscillator as an end mirror, can exhibit nonlinear behaviors [73] which adds complexity to the experiment. The calculations revealed that a power transfer coefficient η of at least 10^3 can be achieved with close to zero mean power at the photodetector in the interferometer readout port. This coefficient is several orders of magnitude higher than the coefficient of the traditional ($\eta \ll 1$) and optical AC coupling ($\eta \approx 1$) schemes.

Chapter 3
Fundamental Limits of Power Stabilization via a Radiation Pressure Transfer Scheme

Abstract In this chapter, the fundamental limits of a power stabilization via a radiation pressure transfer scheme are analyzed. This scheme was chosen based on the results of Chap. 2, which showed that a large transfer coefficient from power to phase modulations can be achieved using a micro-oscillator mirror with low mass and low longitudinal resonance frequency. The induced modulations in the oscillator position, and therefore in the phase of its reflected light, are detected with a Michelson interferometer whose output signal is fed back to a laser power actuator. The interferometer represents the in-loop sensor of this stabilization scheme and therefore interferometer noise sources are imprinted on the out-of-loop beam and set a lower limit to the final power stability. The goal of this chapter is to determine this value for an interferometer fundamentally limited by quantum and thermal noise. Requirements regarding technical noise sources are also discussed. Under the assumption of realistic experimental parameters, the calculations show that generation of a bright squeezed beam is possible.

3.1 Quantum Noise Limit

3.1.1 Mathematical Framework

Let us first introduce the mathematical framework used to calculate the quantum limit of the proposed scheme. The framework is based on the two-photon formalism, developed by Caves and Schumaker [93, 94]. A detailed explanation of this formalism can be found in many references such as the review paper [95], which is the main reference adopted in this thesis, and also [96].

In the two-photon formalism, the electric field of a linearly-polarized and quasi-mono-chromatic wave propagating in vacuum and in the positive direction of the x-axis is re-written using the so called amplitude (c-subscript) and phase (s-subscript) quadratures as basis vectors for the electric field:

M. Trad Nery, *Laser Power Stabilization via Radiation Pressure*, Springer Theses,
https://doi.org/10.1007/978-3-030-95868-8_3

$$\begin{aligned} E(t) &= (\mathcal{E}_0 + e_n(t)) \cos(\omega_0 t + \phi(t) + \phi_0) \\ &\equiv (\mathcal{E}_c + e_c(t)) \cos \omega_0 t + (\mathcal{E}_s + e_s(t)) \sin \omega_0 t , \end{aligned} \tag{3.1}$$

where $\mathcal{E}_0 = \sqrt{8\pi \overline{P}/c\mathcal{A}}$ is the amplitude of the wave in Gaussian-cgs units, ϕ_0 is the wave phase, e_n and ϕ are the amplitude and phase fluctuations respectively, and $\mathcal{A}$ is the effective cross-section area of the light beam.

The use of quadratures as basis vectors is very convenient since they are linearly transformed by the optical devices of interest in this thesis, which is not true for the traditional amplitude-phase description, since it is nonlinear in phase. Another advantage is that quadratures with different frequencies propagate independently from each other. The quadrature coefficients can be derived using the angle sum theorem and assuming small phase fluctuations $\phi(t)$. This leads to the following time-independent terms:

$$\mathcal{E}_c = \mathcal{E}_0 \cos\phi_0 \text{ and } \mathcal{E}_s = -\mathcal{E}_0 \sin\phi_0, \tag{3.2}$$

and time-dependent terms:

$$e_c(t) = e_n(t) \cos\phi_0 - \mathcal{E}_0 \sin\phi_0 \cdot \phi(t) , \tag{3.3}$$

$$e_s(t) = -e_n(t) \sin\phi_0 - \mathcal{E}_0 \cos\phi_0 \cdot \phi(t) , \tag{3.4}$$

where terms containing $e_n(t) \cdot \phi(t)$ were neglected. Without loss of generality, the calculations in this thesis are done assuming that the laser carrier $\mathcal{E}_0$ lies completely in the amplitude quadrature, i.e., $\phi_0 = 0$. This assumption simplifies the mathematics, since amplitude fluctuations $e_n(t)$ will only couple in the amplitude quadrature $e_c(t)$, while phase fluctuations $\phi(t)$ will only couple in the phase quadrature $e_s(t)$. In this case for example, an ideal modulator that acts only on the amplitude of the light will modify only the amplitude quadrature of its transmitted field.

From the quantum mechanics point of view, the electric field Heisenberg operator (in Gaussian-cgs units) as a function of the quadratures can be written as:

$$\hat{E}(t) = \sqrt{\frac{4\pi\hbar\omega_0}{\mathcal{A}c}} \left[\left(A_c + \hat{a}_c(t)\right) \cos\omega_0 t + \left(A_s + \hat{a}_s(t)\right) \sin\omega_0 t \right], \tag{3.5}$$

where A_c and A_s represents the time-independent carrier amplitude and phase quadratures respectively, and $\hat{a}_c(t)$ and $\hat{a}_s(t)$ represent the field fluctuations which can be from classical or quantum nature. In the time domain, quadrature operators have units of $s^{-1/2}$.

In the frequency domain, the quadratures can be written in terms of the creation $\hat{a}$ and anihilation $\hat{a}^\dagger$ operators as:

$$\hat{a}_c(\Omega) = \frac{\hat{a}(\omega_0 + \Omega) + \hat{a}^\dagger(\omega_0 - \Omega)}{\sqrt{2}} \text{ and } \hat{a}_s(\Omega) = \frac{\hat{a}(\omega_0 + \Omega) - \hat{a}^\dagger(\omega_0 - \Omega)}{i\sqrt{2}} . \tag{3.6}$$

As expected, these operators satisfy the usual commutation relations for bosonic fields in the frequency domain:

$$\left[\hat{a}_{\mathrm{c}}(\Omega), \hat{a}_{\mathrm{s}}(\Omega')\right] = 2\pi i \delta(\Omega + \Omega') \quad \text{and} \quad \left[\hat{a}_{\mathrm{c}}(\Omega), \hat{a}_{\mathrm{c}}(\Omega')\right] = \left[\hat{a}_{\mathrm{s}}(\Omega), \hat{a}_{\mathrm{s}}(\Omega')\right] = 0, \tag{3.7}$$

and in the time domain:

$$\left[\hat{a}_{\mathrm{c}}(t), \hat{a}_{\mathrm{s}}(t')\right] = i\delta(t - t') \quad \text{and} \quad \left[\hat{a}_{\mathrm{c}}(t), \hat{a}_{\mathrm{c}}(t')\right] = \left[\hat{a}_{\mathrm{s}}(t), \hat{a}_{\mathrm{s}}(t')\right] = 0. \tag{3.8}$$

The quadratures have to satisfy the Schrödinger uncertainty relation which, in the time domain and for $t = t'$, is given by:

$$\Delta\hat{a}_{\mathrm{c}}^2\, \Delta\hat{a}_{\mathrm{s}}^2 - \left|\frac{\langle\{\hat{a}_{\mathrm{c}}, \hat{a}_{\mathrm{s}}\}\rangle}{2}\right|^2 \geq \left|\frac{\langle[\hat{a}_{\mathrm{c}}, \hat{a}_{\mathrm{s}}]\rangle}{2i}\right|^2 = \frac{1}{4}, \tag{3.9}$$

where $\Delta\hat{a}_{\mathrm{c}}^2$ and $\Delta\hat{a}_{\mathrm{s}}^2$ are the variances of the amplitude and phase quadratures respectively. For a pure state, $\langle\{\hat{a}_{\mathrm{c}}, \hat{a}_{\mathrm{s}}\}\rangle = 0$ and the uncertainty relation is reduced to the known Heisenberg equation:

$$\Delta\hat{a}_{\mathrm{c}}^2\, \Delta\hat{a}_{\mathrm{s}}^2 \geq \frac{1}{4}. \tag{3.10}$$

Therefore, the variances for a state with minimum uncertainty, called coherent state, are $\Delta\hat{a}_{\mathrm{c}}^2 = \Delta\hat{a}_{\mathrm{s}}^2 = 1/2\,\mathrm{s}^{-1}$.

Power Spectral Density

Since the calculations in this thesis are made assuming a linear regime, it is much simpler to analyze the fluctuations/noise in the frequency domain. In this case the figure of merit is the power (or amplitude) spectral density, which expresses the noise or signal strength at a certain Fourier frequency Ω. The power spectral density S_{ij} is defined as the Fourier transform of the correlation function C_{ij} between two observables $\hat{a}_i$ and $\hat{a}_j$. The correlation function is given by the expectation value of the symmetrized product of the operators at different times:

$$C_{ij}(t) = \frac{1}{2}\langle\{\hat{a}_i(t' + t), \hat{a}_j(t')\}\rangle, \quad \left[C_{ij}(t)\right] = \mathrm{s}^{-1}. \tag{3.11}$$

By rewriting this equation in the Fourier domain, the relation for the power spectral density is obtained:

$$\pi S_{ij}(\Omega)\,\delta(\Omega + \Omega') = \frac{1}{2}\langle\{\hat{a}_i(\Omega), \hat{a}_j(\Omega')\}\rangle, \quad \left[S_{ij}\right] = \mathrm{s}^{-1}\cdot\mathrm{Hz}^{-1}. \tag{3.12}$$

Here the integration was done only for positive frequencies and therefore $S_{ij}(\Omega)$ represents the single-sided power spectral density, which is the convention adopted in this thesis.

The Heisenberg uncertainty relation in the frequency domain is obtained by the Fourier transform of Equation 3.9, which leads to:

$$H^2 = S_{\mathrm{cc}} S_{\mathrm{ss}} - |S_{\mathrm{cs}}|^2 \geq 1\,. \tag{3.13}$$

The power and cross power spectral densities for a coherent state are frequency independent and equal to:

$$S_{\mathrm{cc}} = S_{\mathrm{ss}} = 1\,\mathrm{s}^{-1}\cdot\mathrm{Hz}^{-1} \quad \text{and} \quad S_{\mathrm{cs}} = S_{\mathrm{sc}} = 0\,. \tag{3.14}$$

Laser Power

For the calculation of the mirror displacement due to radiation pressure, it is important to express the laser power as a function of the amplitude quadrature operator. The laser power operator can be derived from the relationship of the mean energy stored in the electric field in a given volume $V = cT\mathcal{A}$ (cylinder with cross-sectional area $\mathcal{A}$ and length cT) per measurement time T:

$$\hat{P}(t) = \lim_{T\to 0} \frac{1}{T} \int_V \frac{\overline{\left|\hat{E}(t)\right|^2}}{4\pi} dV\,, \tag{3.15}$$

which was obtained from the Poynting vector equation in Gaussian-cgs units. Here the time average in the electric field is an average over the optical period of the laser. Then, using Eq. 3.5 the following expression can be obtained:

$$\begin{aligned} \hat{P}(t) &= \frac{\hbar\omega_0}{2}\left(A_{\mathrm{c}} + \hat{a}_{\mathrm{c}}(t)\right)^2 + \frac{\hbar\omega_0}{2}\hat{a}_{\mathrm{s}}(t)^2 \\ &\simeq \frac{\hbar\omega_0 A_{\mathrm{c}}^2}{2} + \hbar\omega_0 A_{\mathrm{c}}\hat{a}_{\mathrm{c}}(t), \end{aligned} \tag{3.16}$$

where the terms containing $\hat{a}_{\mathrm{c}}^2(t)$ and $\hat{a}_{\mathrm{s}}^2(t)$ were neglected. The mean laser power $\overline{P}$ and the carrier amplitude A_{c} are then:

$$\overline{P} = \left\langle \hat{P}(t) \right\rangle = \frac{\hbar\omega_0 A_{\mathrm{c}}^2}{2} \;\Rightarrow\; A_{\mathrm{c}} = \sqrt{\frac{2\overline{P}}{\hbar\omega_0}}, \quad [A_{\mathrm{c}}] = \mathrm{Hz}^{-1/2}\,. \tag{3.17}$$

The amplitude spectral density of the laser power fluctuations $\delta P(\Omega)$ is given as a function of the power spectral density S_{cc} of the amplitude quadrature as:

$$\delta P(\Omega) = \sqrt{2\hbar\omega_0 \overline{P}\, S_{\rm cc}}\,, \quad [\delta P(\Omega)] = \mathrm{W}\cdot\mathrm{Hz}^{-1/2}\,. \tag{3.18}$$

The absolute shot noise SN of the laser is calculated considering a coherent state, which simply leads to $\mathrm{SN} = \sqrt{2\hbar\omega_0 \overline{P}}$.

Finally, the amplitude spectral density of the relative power noise is given by:

$$\mathrm{RPN} = \frac{\delta P(\Omega)}{\overline{P}} = \sqrt{\frac{2\hbar\omega_0\, S_{\rm cc}}{\overline{P}}}\,, \quad [\mathrm{RPN}] = \mathrm{Hz}^{-1/2}\,, \tag{3.19}$$

and therefore the relative laser shot noise is $\mathrm{RSN} = \sqrt{2\hbar\omega_0/\overline{P}}$.

Mirror Displacement Due to Radiation Pressure

As seen in Sect. 2.1.3, the mirror displacement $x(\Omega)$ caused by radiation pressure is given by:

$$x(\Omega) = \frac{2P(\Omega)}{c}\cdot \chi_m\,, \tag{3.20}$$

where χ_m is the mechanical susceptibility of the mirror which, in the frequency domain, is:

$$\chi_m\,(\Omega) = \frac{1}{-m\Omega^2 + m\Omega_0^2(1+i\phi)}\,, \quad [\chi_m] = \mathrm{m}\cdot\mathrm{N}^{-1}\,, \tag{3.21}$$

$$|\chi_m\,(\Omega)| = \frac{1}{m\sqrt{\left(\Omega_0^2 - \Omega^2\right)^2 + \Omega_0^4\phi^2}}\,. \tag{3.22}$$

By invoking Eq. 3.16, the time (or frequency)-dependent position operator can be obtained as a function of the amplitude quadrature fluctuations as:

$$\hat{x}(\Omega) = \frac{2\chi_m\sqrt{2\overline{P}\hbar\omega_0}}{c}\cdot \hat{a}_{\rm c}(\Omega) \equiv \alpha\,\chi_m\cdot \hat{a}_{\rm c}(\Omega)\,, \tag{3.23}$$

where $\alpha = \sqrt{8\overline{P}\hbar\omega_0}/c$. Note that since $\hat{a}_{\rm c}$ is a hermitian operator, $\hat{x}$ is immediately hermitian for a viscous loss angle. However, for a structural loss angle, a substitution of $\mathrm{sign}(\Omega)i\phi$ needs to be made in the loss angle term, which then implies $\chi_m^*(\Omega) = \chi_m(-\Omega)$, guaranteeing that $\hat{x}$ remains hermitian.

Field Reflected from a Moving Mirror

Consider an incident field $\hat{b}$ that impinges on a 100 % reflective free moving mirror with position $\hat{x}$, as shown in Fig. 3.1. By choosing the initial phase of the incident

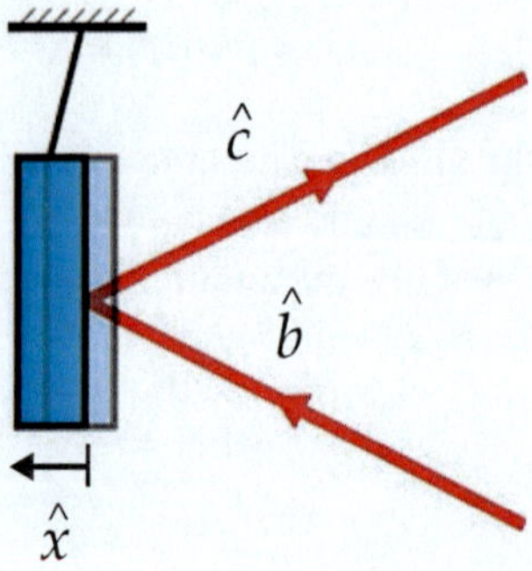

Fig. 3.1 Scheme of an incident light field $\hat{b}$ that is reflected from a moving mirror with position $\hat{x}$, into the field $\hat{c}$. The convention adopted as the positive direction of $\hat{x}$ is shown by the arrow below the mirror

field ϕ_0 to be zero, the carrier will be located solely in the amplitude quadrature, with a magnitude given by Eq. 3.17. The reflected field $\hat{c}$ is affected with a phase shift of $2\omega_0\hat{x}(\Omega)/c$ caused by the moving mirror and, according to Eqs. 3.3, 3.4 and 3.5 is given by:

$$\hat{c}_c(\Omega) = \hat{b}_c(\Omega)\,,$$
$$\hat{c}_s(\Omega) = \hat{b}_s(\Omega) - \sqrt{\frac{8\overline{P}\omega_0}{\hbar c^2}}\cdot\hat{x}(\Omega) = \hat{b}_s(\Omega) - \frac{\alpha}{\hbar}\cdot\hat{x}(\Omega)\,. \tag{3.24}$$

Here it was assumed that the mirror displacement is much smaller than λ_0. This equation shows that, for small displacement, the amplitude quadrature is unchanged upon reflection. The phase quadrature however, acquires an additional positive phase for a mirror moving in the positive direction.

If the mirror is displaced due to radiation pressure from the incident field $\hat{b}$, then, according to Eq. 3.23, the reflected phase quadrature is given by:

$$\hat{c}_s(\Omega) = \hat{b}_s(\Omega) - \mathcal{K}_\alpha\cdot\hat{b}_c(\Omega) \tag{3.25}$$

where $|\mathcal{K}_\alpha| = \alpha^2\,|\chi_m|\,/\hbar$ expresses the magnitude of the transfer function from amplitude quadrature of the incident field to phase quadrature of the reflected field.

Beamsplitter Matrix

The field relations for a static and lossless beamsplitter with transmission and reflection amplitude coefficients of $\sqrt{T}$ and $\sqrt{R}$ (see Fig. 3.2), can be written as:

$$\begin{pmatrix}\hat{h}\\ \hat{c}\\ \hat{a}\\ \hat{r}\end{pmatrix} = \begin{pmatrix}-\sqrt{R} & 0 & 0 & \sqrt{T}\\ 0 & -\sqrt{R} & \sqrt{T} & 0\\ 0 & \sqrt{T} & \sqrt{R} & 0\\ \sqrt{T} & 0 & 0 & \sqrt{R}\end{pmatrix}\begin{pmatrix}\hat{d}\\ \hat{v}\\ \hat{s}\\ \hat{b}\end{pmatrix}\,. \tag{3.26}$$

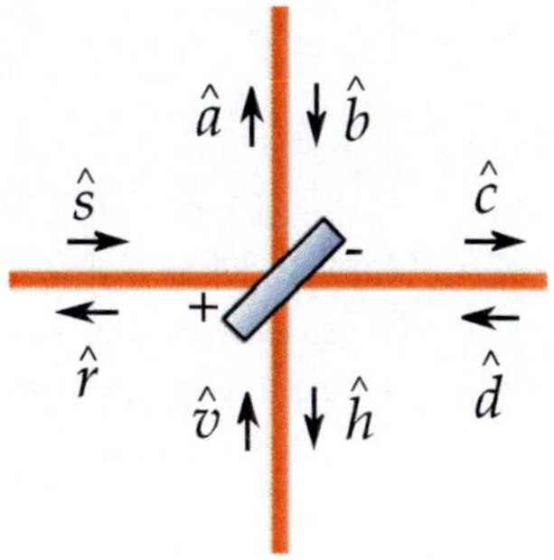

Fig. 3.2 Definition of the fields used to derive the beamsplitter matrix. The convention for the positive and negative signs acquired during reflection is shown

3.1.2 Traditional Scheme

First, let us calculate the quantum limit of the traditional power stabilization scheme introduced in Sect. 1.6.1 using the described two-photon formalism approach. The field operators used in this calculation are represented in Fig. 3.3. Using the beam-splitter matrix of Equation 3.26, the amplitude and phase quadratures of the out-of-loop field $\hat{o}$ are obtained:

$$\hat{o} = \sqrt{T} \cdot \hat{t} - \sqrt{R} \cdot \hat{v} \,. \tag{3.27}$$

The quadratures for the field $\hat{t}$ transmitted by the amplitude modulator are:

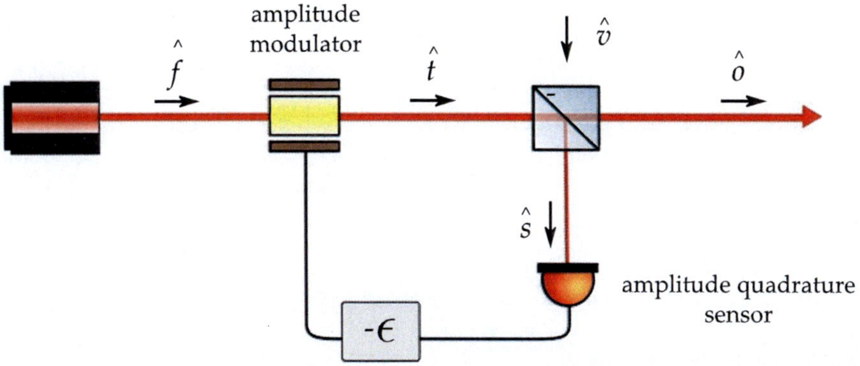

Fig. 3.3 Schematic of the traditional amplitude stabilization scheme. The free running laser field $\hat{f}$ is sent to an amplitude modulator, and the transmitted field $\hat{t}$ is partially reflected by a beamsplitter and partially transmitted to the out-of-loop field $\hat{o}$. The vacuum field $\hat{v}$ couples at the open port of the beamsplitter and is partially imprinted on the $\hat{t}$ field by the feedback control loop, and partially reflected to the $\hat{o}$ field. The in-loop amplitude quadrature $\hat{s}_c$ is sensed and the resulting feedback signal is amplified by a controller with a complex gain ϵ

$$\hat{\imath}_{\rm c} = \hat{f}_{\rm c} - \hat{s}_{\rm c} \cdot \epsilon, \tag{3.28}$$

$$\hat{\imath}_{\rm s} = \hat{f}_{\rm s}, \tag{3.29}$$

where ϵ is the complex amplification factor of the control loop. Here an ideal amplitude modulator, as discussed in Sect. 3.1.1, was assumed. The field $\hat{s}$ is a sum of the reflected field $\hat{\imath}$ from the beamsplitter and the transmitted vacuum field $\hat{v}$ coupling at the open port of the beamsplitter:

$$\hat{s} = \sqrt{T} \cdot \hat{v} + \sqrt{R} \cdot \hat{\imath}\,. \tag{3.30}$$

According to a classical feedback control theory, Eqs. 3.28 and 3.30 can be combined in the steady state regime and therefore the final expression for $\hat{\imath}_{\rm c}$ is:

$$\hat{\imath}_{\rm c} = \frac{\hat{f}_{\rm c} - \epsilon\sqrt{T} \cdot \hat{v}_{\rm c}}{1 + \epsilon\sqrt{R}}\,. \tag{3.31}$$

The out-of-loop field quadratures as a function of the uncorrelated input fields are then:

$$\hat{o}_{\rm c} = \frac{\sqrt{T} \cdot \hat{f}_{\rm c} - \left(\epsilon + \sqrt{R}\right) \cdot \hat{v}_{\rm c}}{1 + \epsilon\sqrt{R}}\,, \tag{3.32}$$

$$\hat{o}_{\rm s} = \sqrt{T} \cdot \hat{f}_{\rm s} - \sqrt{R} \cdot \hat{v}_{\rm s}\,. \tag{3.33}$$

The free running amplitude quadrature $\hat{f}_{\rm c}$ is reduced by the open loop gain $\epsilon\sqrt{R}$, that depends on the transfer function $\mathcal{S} = \sqrt{R}$ from amplitude modulations at the output of the modulator to amplitude modulations at the photodetector.[1] The vacuum field is partially suppressed by the loop gain and partially imprinted on the out-of-loop field.

It is an interesting fact that, while the out-of-loop field satisfies the expected bosonic commutation relation:

$$\left[\hat{o}_{\rm c}(\Omega)\,, \hat{o}_{\rm s}(\Omega')\right] = 2\pi i\delta\left(\Omega + \Omega'\right) \cdot \frac{T + \epsilon\sqrt{R} + R}{1 + \epsilon\sqrt{R}} = 2\pi i\delta\left(\Omega + \Omega'\right)\,, \tag{3.34}$$

the in-loop field does not:

$$\left[\hat{\imath}_{\rm c}(\Omega)\,, \hat{\imath}_{\rm s}(\Omega')\right] = \frac{2\pi i\delta\left(\Omega + \Omega'\right)}{1 + \epsilon\sqrt{R}} \neq 2\pi i\delta\left(\Omega + \Omega'\right)\,, \tag{3.35}$$

which violates the uncertainty principle. Such violation comes from the assumption that the modulator acts only on the amplitude quadrature of the laser, which is

[1] The transfer function $\mathcal{S}$ is the amplitude analogous to the power η coefficient introduced in Chap. 2.

suppressed, while the phase quadrature is unchanged. A solution for this problem was proposed in reference [97], and is called self-consistent formalism. There the modulator is described via the squeezing operator, which introduces anti-squeezing in the phase quadrature, preserving the commutation relation for the in-loop field $\hat{t}$. However, to preserve the relations for the out-of-loop field, the modulator needs to act on quadratures ($\hat{\upsilon}_s$ in this case) which are never physically present in the modulator, violating the causality principle. In this thesis, the violation for the in-loop field $\hat{t}$ was chosen not to be taken into account since, to my knowledge, it is not possible to measure $\hat{t}$ without coupling additional noise in the measurement device, which should force the preservation of the commutation relation.

In the limit when $|\epsilon|\sqrt{R} \gg 1$ and subsequently when $|\epsilon| \to \infty$, the amplitude quadrature and its corresponding power spectral density are:

$$\hat{o}_c \simeq \frac{-\hat{\upsilon}_c}{\sqrt{R}} \quad \text{and} \quad S^{\hat{o}}_{cc} \simeq \frac{1}{R}. \tag{3.36}$$

The cross power spectral density S_{cs} is always zero, and this is due to the fact that the scheme does not introduce a correlation between the amplitude and phase quadratures. However the uncertainty relation between the quadratures is higher than for a coherent state, which is expected since noise is always imprinted in the amplitude quadrature in comparison to a vacuum state ($R < 1$). According to Eq. 3.18, the ASD of the out-of-loop absolute power noise is:

$$\delta P_{ool} = \sqrt{\frac{2\hbar\omega_0 \overline{P_0}\,(1-R)}{R}}, \tag{3.37}$$

where $\overline{P_0}$ is the mean power at the output of the laser. The relative power noise is then:

$$\mathrm{RPN}_{ool} = \sqrt{\frac{2\hbar\omega_0}{\overline{P_0}\,R(1-R)}}, \tag{3.38}$$

which is the same result as obtained in Eq. 1.9.

If one is interested in calculating the performance of the traditional scheme assisted by squeezing, the power spectral density of the operator $\hat{\upsilon}_c$ has to be substituted to its squeezed state value, which is smaller than $1\ \mathrm{s}^{-1}\cdot\mathrm{Hz}^{-1}$ (see Sect. 1.6.2).

3.1.3 Radiation Pressure Scheme

The schematic of the stabilization scheme via radiation pressure chosen to be analyzed in this thesis is shown in Fig. 3.4. This scheme implements two independent laser sources: one providing a strong transfer beam to be stabilized (red trace), and

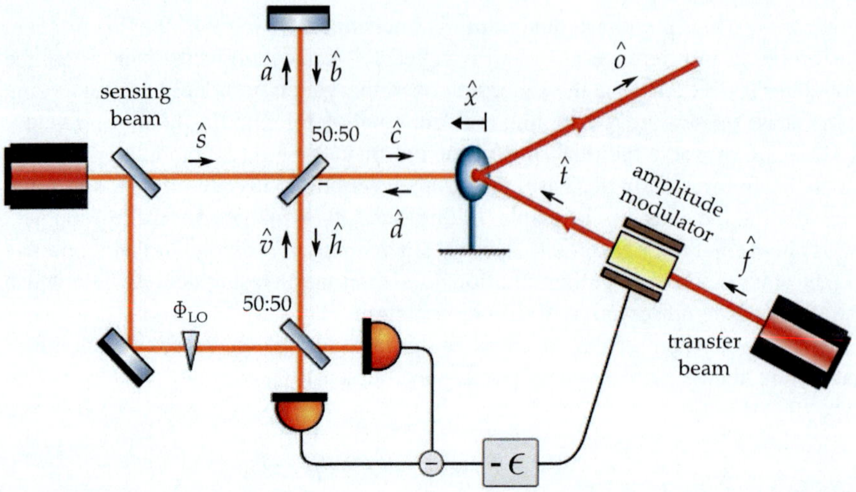

Fig. 3.4 Schematic of the amplitude stabilization scheme via radiation pressure. The free running laser field $\hat{f}$ is sent to an amplitude modulator and the transmitted field $\hat{t}$ impinges on a micro-oscillator mirror and is fully reflected to the out-of-loop field $\hat{o}$. The oscillator's position is sensed via a Michelson interferometer with a balanced homodyne readout. The feedback signal is amplified by a controller with a complex gain ϵ. The convention adopted as the positive direction of $\hat{x}$ is shown by the arrow on top of the oscillator

other a weak sensing beam (orange trace), that performs the readout of the position of the micro-oscillator mirror.[2] Parts of this section are published in [98].

The scheme works as follows: the free running field $\hat{f}$ to be stabilized from the transfer beam is sent to an amplitude modulator, and the transmitted field $\hat{t}$ impinges on a micro-oscillator mirror and is fully reflected to the out-of-loop field $\hat{o}$. The oscillator's position $\hat{x}$ is sensed via a Michelson interferometer with a balanced homodyne readout [95] ideally sensitive only to the phase quadrature of the output field $\hat{h}$, that represents the differential length of the interferometer arms. The output signal is then amplified by a complex gain ϵ, and the resulting feedback signal is sent to the amplitude modulator, which stabilizes $\hat{f}_c$. The interferometer implements a weak sensing beam represented by the field $\hat{s}$. The figure also shows the vacuum field $\hat{v}$ that couples at the dark port of the interferometer. A similar scheme was analyzed in reference [99], but for a freely suspended mirror inside an optical cavity. First, let us introduce the following useful parameters, similarly defined in Eqs. 3.25 and 3.23:

$$
\begin{aligned}
\mathcal{K}'_\alpha &= \frac{\alpha'^2 \chi_m}{\hbar}, \quad \text{where } \alpha' = \frac{\sqrt{8 P_s \hbar \omega_0}}{c}, \\
\mathcal{K}_\beta &= \frac{\beta^2 \chi_m}{\hbar}, \quad \text{where } \beta = \frac{\sqrt{8 P_t \hbar \omega_0}}{c},
\end{aligned}
\tag{3.39}
$$

[2] Here oscillator is referred to the combination of the mirror plus its suspension.

where P_s and P_t are the mean power of the sensing and transfer beams respectively (the overline was omitted for simplicity). Then, using the beamsplitter matrix from Eq. 3.26, and assuming that the interferometer is locked on the dark fringe and has equal arm lengths, the relations for the fields in the interferometer are obtained:

$$\begin{aligned}
\hat{\boldsymbol{a}} &= \sqrt{R}\,\hat{\boldsymbol{s}} + \sqrt{T}\,\hat{\boldsymbol{v}}\,,\\
\hat{\boldsymbol{b}} &= -\hat{\boldsymbol{a}},\\
\hat{\boldsymbol{c}} &= \sqrt{T}\,\hat{\boldsymbol{s}} - \sqrt{R}\,\hat{\boldsymbol{v}}\,,\\
\hat{\boldsymbol{d}} &= -\hat{\boldsymbol{c}} - \begin{pmatrix}0\\1\end{pmatrix}\frac{\sqrt{T}\alpha'}{\hbar}\hat{x}\,,\\
\hat{\boldsymbol{h}} &= \sqrt{T}\,\hat{\boldsymbol{b}} - \sqrt{R}\,\hat{\boldsymbol{d}}\,,
\end{aligned} \tag{3.40}$$

where the operators are represented in the vector form: $\hat{\boldsymbol{a}} = \left(\hat{a}_c\ \hat{a}_s\right)^T$ and optical losses were neglected. The oscillator's position is influenced by the two independent fields $\hat{s}$ and $\hat{v}$ in the interferometer, and by the stabilized field $\hat{t}$:

$$\hat{x} = \chi_m\left(-\alpha' T\cdot\hat{s}_c + \alpha'\sqrt{RT}\cdot\hat{v}_c + \beta\cdot\hat{t}_c\right). \tag{3.41}$$

By implementing an interferometer with a balanced homodyne readout, the control loop feedback signal is proportional to:

$$\hat{h}_s\sin\Phi_{LO} + \hat{h}_c\cos\Phi_{LO}\,, \tag{3.42}$$

where Φ_{LO} is the homodyne angle between the local oscillator and the interferometer output. In order to improve the signal-to-noise ratio, a readout only sensitive to the phase quadrature was chosen, i.e. $\Phi_{LO} = \pi/2$. Additionally, noise sources from the local oscillator beam ideally do not couple at the readout. Accordingly to Eq. 3.40, the phase quadrature in the output of the interferometer is:

$$\hat{h}_s = -\hat{v}_s - \mathcal{K}'_\alpha T\sqrt{TR}\cdot\hat{s}_c + \mathcal{K}'_\alpha TR\cdot\hat{v}_c + \sqrt{TR\mathcal{K}'_\alpha\mathcal{K}_\beta}\cdot\hat{t}_c\,. \tag{3.43}$$

Assuming a 50:50 beamsplitter and defining $\mathcal{K}_\alpha = \mathcal{K}'_\alpha/4$, this equation simplifies to:

$$\hat{h}_s = -\hat{v}_s - \mathcal{K}_\alpha\cdot\left(\hat{s}_c - \hat{v}_c\right) + \sqrt{\mathcal{K}_\alpha\mathcal{K}_\beta}\cdot\hat{t}_c\,. \tag{3.44}$$

The first term represents the vacuum fluctuations $\hat{v}_s$ in the phase quadrature that couples via the dark port of the interferometer and are fully reflected to the output field. They are referred here as the interferometer readout noise. The second term represents the radiation pressure noise caused by the amplitude quadratures $\hat{s}_c$ and $\hat{v}_c$ of the sensing and vacuum fields. The third term contains information about the free running amplitude quadrature $\hat{f}_c$, and therefore is a representation of the interested signal. The factor $\sqrt{\mathcal{K}_\alpha \mathcal{K}_\beta} = \alpha\beta\chi_m/\hbar$ represents the transfer function S from amplitude modulations from the transfer beam to amplitude modulations detected by the in-loop photodetector.

Similarly to the calculation for the traditional scheme, the amplitude quadrature transmitted by the modulator is given by $\hat{t}_c = \hat{f}_c - \epsilon \cdot \hat{h}_s$ which, in the steady state regime, can be combined with Eq. 3.44 to obtain:

$$\begin{aligned}\hat{t}_c &= \frac{\hat{f}_c + \epsilon\left(\mathcal{K}_\alpha \cdot (\hat{s}_c - \hat{v}_c) + \hat{v}_s\right)}{1 + \epsilon\sqrt{\mathcal{K}_\alpha \mathcal{K}_\beta}}\,, \\ \hat{t}_s &= \hat{f}_s\,,\end{aligned} \tag{3.45}$$

where an ideal modulator acting only on the amplitude quadrature of the light was considered. For this calculation it was assumed that the negative feedback does not introduce a new quantum noise source in the stabilized field $\hat{t}$. This assumption is sustained by the fact that, because the control loop performs a single quadrature measurement of the photon-number operator, the commutation relation for the amplified field is preserved without the need of an additional noise operator [100].

The out-of-loop quadratures are obtained from Eqs. 3.24 and 3.41:

$$\begin{aligned}\hat{o}_c &= \hat{t}_c\,, \\ \hat{o}_s &= \hat{t}_s - \frac{\beta}{\hbar}\cdot\hat{x} = \hat{t}_s + \sqrt{\mathcal{K}_\alpha \mathcal{K}_\beta}\cdot(\hat{s}_c - \hat{v}_c) - \mathcal{K}_\beta \cdot \hat{t}_c\,,\end{aligned} \tag{3.46}$$

which, according to Eq. 3.45, are given as functions of the uncorrelated input fields as:

$$\hat{o}_c = \frac{\hat{f}_c + \epsilon\cdot\hat{v}_s + \epsilon\mathcal{K}_\alpha\cdot(\hat{s}_c - \hat{v}_c)}{1 + \epsilon\sqrt{\mathcal{K}_\alpha \mathcal{K}_\beta}} \tag{3.47}$$

$$\approx \frac{\hat{v}_s}{\sqrt{\mathcal{K}_\alpha \mathcal{K}_\beta}} + \sqrt{\frac{\mathcal{K}_\alpha}{\mathcal{K}_\beta}}\cdot(\hat{s}_c - \hat{v}_c)\,, \tag{3.48}$$

$$\hat{o}_s = \hat{f}_s - \frac{\mathcal{K}_\beta\cdot\hat{f}_c + \epsilon\mathcal{K}_\beta\cdot\hat{v}_s - \sqrt{\mathcal{K}_\alpha \mathcal{K}_\beta}\cdot(\hat{s}_c - \hat{v}_c)}{1 + \epsilon\sqrt{\mathcal{K}_\alpha \mathcal{K}_\beta}} \tag{3.49}$$

$$\approx \hat{f}_s - \hat{v}_s\sqrt{\frac{\mathcal{K}_\beta}{\mathcal{K}_\alpha}}\,. \tag{3.50}$$

The approximation was done in the high gain regime when $\left|\epsilon\sqrt{\mathcal{K}_\alpha\mathcal{K}_\beta}\right| \gg 1$ and subsequently $|\epsilon| \to \infty$. As expected, the free running modulations $\hat{f}_c$ in the out-of-loop amplitude quadrature are reduced by increasing the open loop gain $\epsilon\sqrt{\mathcal{K}_\alpha\mathcal{K}_\beta}$ of the feedback control loop. Additionally, the readout and radiation pressure noise introduced by the interferometer (second and third terms in Eq. 3.47) are imprinted on the out-of-loop field and cannot be decreased below a certain value (Eq. 3.48).

By comparing Eqs. 3.36 and 3.48 a remarkable difference between the traditional and the radiation pressure schemes can be observed. In the traditional scheme, the sensor noise imprinted on the out-of-loop beam is divided by $\mathcal{S} = \sqrt{R}$ which is always smaller than 1 for a linear beam splitter. However, in the radiation pressure scheme, the interferometer readout noise is divided by $\mathcal{S} = \sqrt{\mathcal{K}_\alpha\mathcal{K}_\beta}$ which, for frequencies below resonance, can have an absolute value much larger than 1 for a high susceptibility oscillator. Additionally, the radiation pressure noise is multiplied by $\sqrt{\mathcal{K}_\alpha/\mathcal{K}_\beta}$, i.e, the ratio between the sensing and transfer beam mean power, which can easily be set to smaller than 1. For these reasons, it is possible to obtain bright squeezed light in the out-of-loop beam, i.e., a power stability below the transfer beam shot noise, which is not possible in the traditional and OAC-coupling schemes.

The single sided power spectral densities of the amplitude ($S^{\hat{o}}_{\mathrm{cc}}$) and phase quadratures ($S^{\hat{o}}_{\mathrm{ss}}$) of the out-of-loop beam are respectively:

$$S^{\hat{o}}_{\mathrm{cc}} = \frac{S^{\hat{f}}_{\mathrm{cc}} + |\epsilon|^2 + |\epsilon\mathcal{K}_\alpha|^2\left(S^{\hat{s}}_{\mathrm{cc}} + 1\right)}{\left|1 + \epsilon\sqrt{\mathcal{K}_\alpha\mathcal{K}_\beta}\right|^2} \approx \frac{1}{\left|\mathcal{K}_\alpha\mathcal{K}_\beta\right|} + \left|\frac{\mathcal{K}_\alpha}{\mathcal{K}_\beta}\right|\left(S^{\hat{s}}_{\mathrm{cc}} + 1\right), \tag{3.51}$$

$$S^{\hat{o}}_{\mathrm{ss}} = S^{\hat{f}}_{\mathrm{ss}} + \frac{\left|\mathcal{K}_\beta\right|^2\left(S^{\hat{f}}_{\mathrm{cc}} + |\epsilon|^2\right) + \left|\mathcal{K}_\alpha\mathcal{K}_\beta\right|\left(S^{\hat{s}}_{\mathrm{cc}} + 1\right)}{\left|1 + \epsilon\sqrt{\mathcal{K}_\alpha\mathcal{K}_\beta}\right|^2} \approx S^{\hat{f}}_{\mathrm{ss}} + \left|\frac{\mathcal{K}_\beta}{\mathcal{K}_\alpha}\right|, \tag{3.52}$$

where the approximations were done for the high gain regime.

3.1.3.1 Analysis for Coherent Beams

To understand the dynamics of this scheme, let us first assume that the sensing and transfer beams are in a coherent state, i.e., $S^{\hat{f}(\hat{s})}_{\mathrm{cc}} = 1$ and $S^{\hat{f}(\hat{s})}_{\mathrm{ss}} = 1$. Figure 3.5 shows the ASD of the amplitude (upper plot) and phase (lower plot) quadratures for different amplification factors ϵ (here treated as frequency independent). The oscillator chosen for the analysis in this chapter is a micro-oscillator with a mass of $m = 40\,\mathrm{ng}$, fundamental longitudinal resonance frequency $\Omega_0 = 2\pi \cdot 100\,\mathrm{Hz}$ and a quality factor $Q = 2 \times 10^5$. More details about this choice are given in Sect. 3.2. Additionally, the wavelengths of the transfer and sensing beams were set to 1064 nm.

First let us analyze the amplitude quadrature. When the loop is open, $|\epsilon| = 0$ and therefore $S^{\hat{o}}_{\mathrm{cc}} = S^{\hat{f}}_{\mathrm{cc}} = 1$ for a coherent state. As anticipated, when the loop is closed an interesting result is obtained: for low frequencies the amplitude quadrature is squeezed, as it has an ASD smaller than for the vacuum state, whereas for high

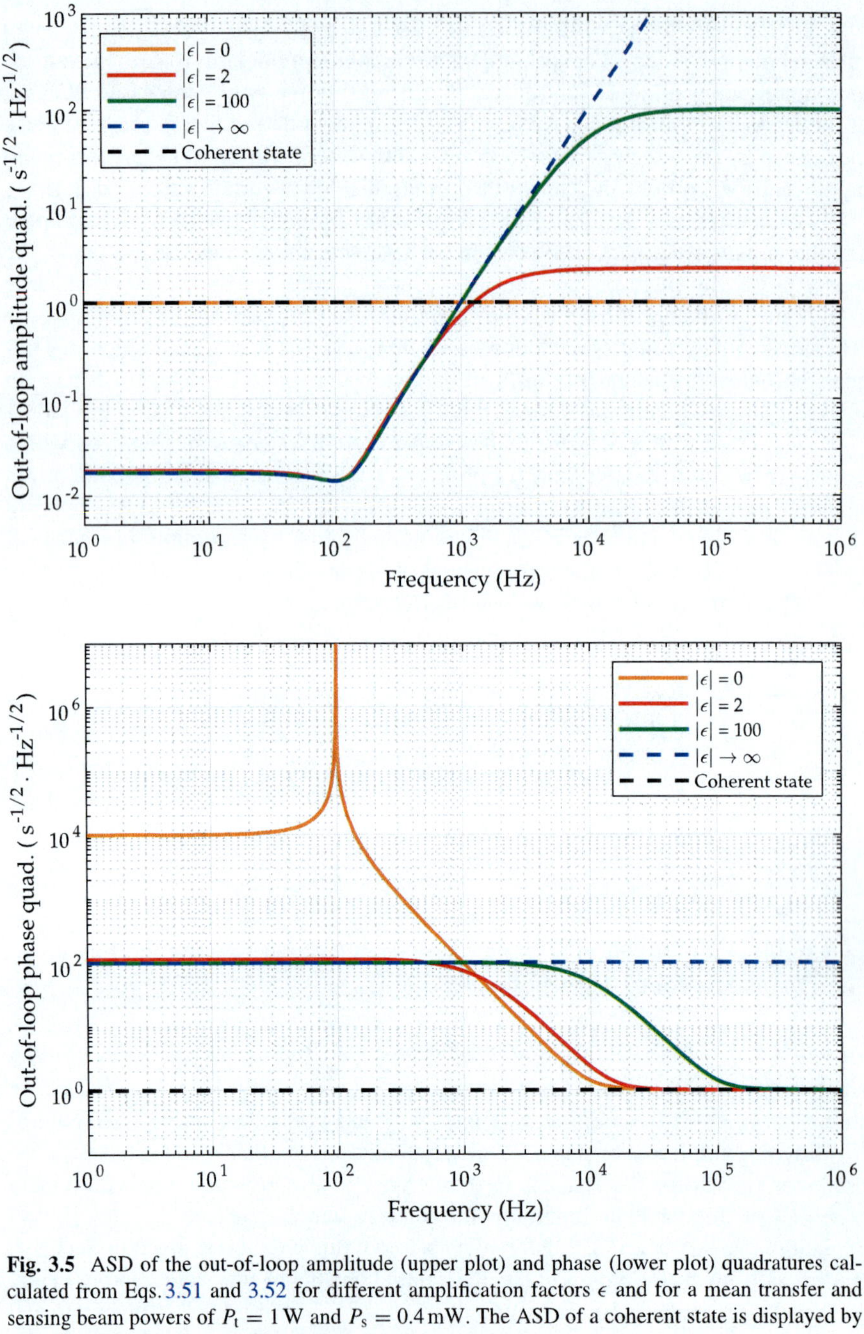

Fig. 3.5 ASD of the out-of-loop amplitude (upper plot) and phase (lower plot) quadratures calculated from Eqs. 3.51 and 3.52 for different amplification factors ϵ and for a mean transfer and sensing beam powers of $P_t = 1\,\mathrm{W}$ and $P_s = 0.4\,\mathrm{mW}$. The ASD of a coherent state is displayed by the dashed black line as a reference

frequencies noise is always imprinted. This can be better understood by analyzing the amplitude quadrature in the high gain regime (Eq. 3.48). In this case, $\hat{f}_c$ is completely suppressed and has no contribution to the spectral density. The interferometer readout noise $\hat{v}_s$ is divided by the oscillator's susceptibility and by the transfer and sensing beam mean powers since $\sqrt{\mathcal{K}_\alpha \mathcal{K}_\beta} \propto \chi_m \sqrt{P_t P_s}$. This means that below Ω_0 the readout noise is suppressed by a constant and large factor ($|\chi_m| \approx 6 \times 10^4 \mathrm{m} \cdot \mathrm{N}^{-1}$), while above Ω_0 the suppression gets smaller and is subsequently amplified in a Ω^2 fashion, following the decrease in the susceptibility proportional to $1/\Omega^2$. This amplification happens because, since the optomechanical response of the oscillator at high frequencies gets smaller, the loop needs to apply a higher amplitude modulation on the transfer beam to compensate the readout noise and keep the error signal zero. In this regime it is desirable to use small or zero feedback amplification ϵ. The power dependence on the readout noise suppression is due to the fact that by increasing P_s the output interferometer signal is increased (higher signal-to-shot-noise ratio), and by increasing P_t the modulation that needs to be applied in the transfer beam to compensate the readout noise is reduced. The contribution from radiation pressure noise (from the quadratures $\hat{v}_c$ and $\hat{s}_c$) does not depend on the mechanical susceptibility and therefore it is frequency independent. This is to be expected since the susceptibility has an equal impact on the radiation pressure displacement caused by $\hat{t}_c$, $\hat{s}_c$, and $\hat{v}_c$. It depends, however, on the ratio between the sensing and transfer beam mean powers, which is smaller than 1.

The Heisenberg uncertainty principle anticipates that in the region where the amplitude quadrature is squeezed, the phase quadrature will suffer a penalty and be anti-squeezed. This is exactly what is shown in the lower plot of Fig. 3.5. When the loop is open, the phase quadrature has contributions from $\hat{f}_s$ (with $S_{ss}^{\hat{f}} = 1$), plus the frequency dependent terms $\mathcal{K}_\beta \cdot \hat{f}_c$ and $\sqrt{\mathcal{K}_\beta \mathcal{K}_\alpha} \cdot (\hat{s}_c - \hat{v}_c)$, representing the radiation pressure displacement caused by the transfer and sensing beams. When the loop is closed, the amplitude quadrature $\hat{f}_c$ is suppressed for low frequencies, which consequently leads to a reduction in the phase quadrature $\hat{o}_s$. However, unlike the amplitude, the phase quadrature will never undergo squeezing since there is no mechanism in the setup to suppress $\hat{f}_s$. In the regime where $|\epsilon| \to \infty$, an additional constant noise is imprinted on the phase quadrature at high frequencies. This is because the noise in the amplitude quadrature $\hat{t}_c$ increases with Ω^2, while the susceptibility decreases with $1/\Omega^2$, resulting in a frequency independent contribution.

Heisenberg Uncertainty Relation

The out-of-loop quadratures satisfy the usual bosonic commutation relation:

$$\begin{aligned}\left[\hat{o}_c(\Omega), \hat{o}_s(\Omega')\right] &= 2\pi i \delta\left(\Omega + \Omega'\right)\left[\frac{1}{1+\epsilon\sqrt{\mathcal{K}_\alpha \mathcal{K}_\beta}} + \frac{\epsilon\sqrt{\mathcal{K}_\alpha \mathcal{K}_\beta} + \epsilon^2 \mathcal{K}_\alpha \mathcal{K}_\beta}{\left(1+\epsilon\sqrt{\mathcal{K}_\alpha \mathcal{K}_\beta}\right)^2}\right] \\ &= 2\pi i \delta\left(\Omega + \Omega'\right) .\end{aligned} \tag{3.53}$$

They also need to satisfy the uncertainty relation H given by Eq. 3.10. The cross correlation between the amplitude and phase quadratures is given by:

$$\frac{\langle\{\hat{o}_c, \hat{o}_s\}\rangle}{2} = -\frac{\mathcal{K}_\beta \cdot \left\langle \hat{f}_c^2 \right\rangle}{\left(1+\epsilon\sqrt{\mathcal{K}_\alpha \mathcal{K}_\beta}\right)^2} - \frac{\mathcal{K}_\beta \epsilon^2 \cdot \langle \hat{v}_s^2 \rangle}{\left(1+\epsilon\sqrt{\mathcal{K}_\alpha \mathcal{K}_\beta}\right)^2} + \frac{\mathcal{K}_\alpha \epsilon \sqrt{\mathcal{K}_\alpha \mathcal{K}_\beta} \cdot \langle \hat{s}_c^2 + \hat{v}_c^2 \rangle}{\left(1+\epsilon\sqrt{\mathcal{K}_\alpha \mathcal{K}_\beta}\right)^2}. \tag{3.54}$$

Therefore the cross power spectral density, considering that the transfer and sensing beams are in a coherent state, is then:

$$S_{cs}^{\hat{o}} = \frac{-\mathcal{K}_\beta - \mathcal{K}_\beta \epsilon^2 + 2\epsilon \mathcal{K}_\alpha \sqrt{\mathcal{K}_\alpha \mathcal{K}_\beta}}{\left(1+\epsilon\sqrt{\mathcal{K}_\alpha \mathcal{K}_\beta}\right)^2} \neq 0. \tag{3.55}$$

As expected, the cross spectral density is non-zero, since the micro-oscillator introduces a correlation between the amplitude and phase quadratures of the out-of-loop field.

Figure 3.6 shows the Heisenberg uncertainty relation H for different conditions. In the absence of the sensing beam, $S_{cs}^{\hat{o}} = \mathcal{K}_\beta$ and Heisenberg's uncertainty is then frequency independent and equal to 1, as expected from a pure coherent state (dashed black line). In the presence of the sensing beam, but with the loop open (yellow curve),

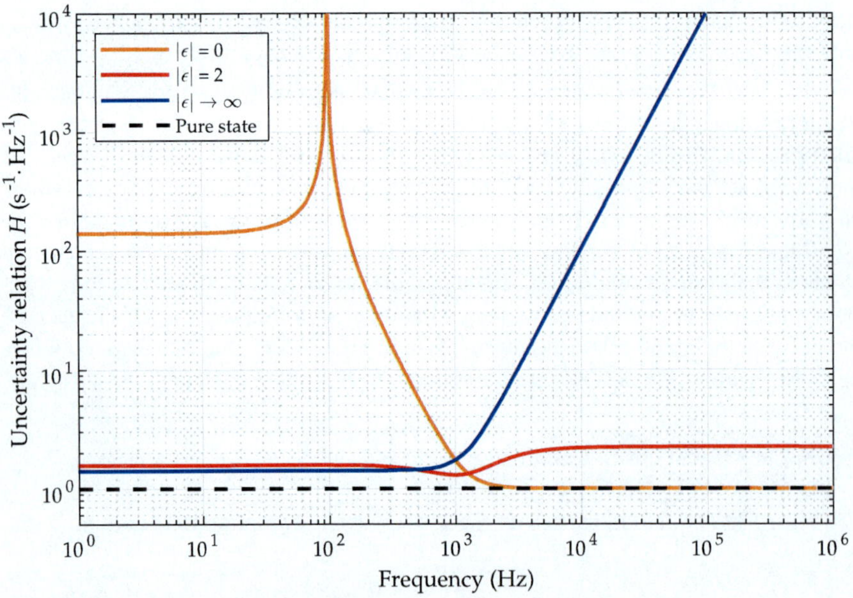

Fig. 3.6 Uncertainty relation H for different amplification factors ϵ in the control loop. For all curves $P_t = 1\,\text{W}$ and $P_s = 0.4\,\text{mW}$. The relation H for a pure coherent state (sensing beam power turned off) is displayed by the dashed black line as a reference

the oscillator moves freely and radiation pressure noise from the interferometer introduces an uncertainty in the out-of-loop phase quadrature that is shaped by the oscillator's susceptibility:

$$H^2 = 1 + 2\left|\mathcal{K}_\alpha \mathcal{K}_\beta\right| . \tag{3.56}$$

When the loop is closed, the position of the oscillator is suppressed at low frequencies, reducing the uncertainty, while for high frequencies an additional uncertainty is introduced in comparison with when the loop is open. The uncertainty principle is always satisfied, which can be easier seen in the regime where $|\epsilon| \to \infty$ (blue curve):

$$H^2 = S^{\hat{o}}_{cc} S^{\hat{o}}_{ss} - \left|S^{\hat{o}}_{cs}\right|^2 = \frac{1}{\left|\mathcal{K}_\alpha \mathcal{K}_\beta\right|} + 2\left|\frac{\mathcal{K}_\alpha}{\mathcal{K}_\beta}\right| + 2 > 1 . \tag{3.57}$$

The uncertainty will always be larger than 1 since noise above the vacuum level is always imprinted in the out-of-loop phase and amplitude quadratures by the interferometer and the control loop.

3.1.3.2 Analysis for Non-coherent Beams

The assumption made that the transfer and sensing beams are in a coherent state was useful to understand the scheme dynamics, but it is unrealistic since lasers always exhibits technical noise. Therefore, in this section the out-of-loop performance is analyzed taking into account technical power noise in both beams.

The power spectral density of the out-of-loop beam amplitude quadrature as a function of the initial relative power noise in the sensing and transfer beams ($\mathrm{RPN_s}$ and $\mathrm{RPN_f}$) can be obtained, in the regime where $\left|\epsilon\sqrt{\mathcal{K}_\alpha \mathcal{K}_\beta}\right| \gg 1$, from Eqs. 3.51 and 3.19:

$$S^{\hat{o}}_{cc} = \frac{c^4}{16 P_s \omega_0^2 \left|\chi_m\right|^2} \left(\frac{\mathrm{RPN_f^2}}{2\hbar\omega_0 \left|\epsilon\right|^2} + \frac{1}{P_t}\right) + \frac{P_s}{4P_t}\left(\frac{\mathrm{RPN_s^2} P_s}{2\hbar\omega_0} + 1\right) . \tag{3.58}$$

In order to obtain squeezing in the out-of-loop amplitude quadrature, $S^{\hat{o}}_{cc}$ needs to be smaller than 1, and therefore the following conditions need to be satisfied:

$$\mathrm{RPN_s} P_s < 2\sqrt{2\hbar\omega_0 P_t} , \tag{3.59}$$

$$\mathrm{RPN_f} < \frac{4\omega_0 \left|\chi_m \epsilon\right| \sqrt{2\hbar\omega_0 P_s}}{c^2} . \tag{3.60}$$

The first condition states that the absolute power noise of the sensing beam needs to be smaller than 2 times the absolute shot noise of the transfer beam. This is to be expected since, for light to be squeezed, the interferometer needs to have a sensitivity high enough to detect the transfer beam shot noise. The second condition is related to the suppression provided by the control loop and can be satisfied in the high open loop gain regime.

The power spectral density of the out-of-loop beam relative power noise $\mathrm{RPN}_{\mathrm{ool,q}}$ is then given by:

$$\mathrm{RPN}^2_{\mathrm{ool,q}} = \frac{c^4}{16\omega_0^2 \left|\chi_m\right|^2} \cdot \frac{\mathrm{RPN}^2_{\mathrm{f}}}{P_{\mathrm{s}} P_{\mathrm{t}} \left|\epsilon\right|^2} + \frac{\mathrm{RPN}^2_{\mathrm{s}} P^2_{\mathrm{s}}}{4P^2_{\mathrm{t}}} + \frac{c^4 \hbar}{8\omega_0 \left|\chi_m\right|^2} \cdot \frac{1}{P_{\mathrm{s}} P^2_{\mathrm{t}}} + \frac{\hbar \omega_0 P_{\mathrm{s}}}{2P^2_{\mathrm{t}}} \,. \tag{3.61}$$

The stability is now limited by the suppression of transfer beam power noise by the control loop (first term), by the technical radiation pressure noise from the sensing beam (second term), and by the interferometer readout and radiation pressure quantum noises (third and fourth terms). The transfer beam mean power has the same effect in all terms: by increasing its value, the $\mathrm{RPN}_{\mathrm{ool,q}}$ is reduced. This happens for two reasons: first because the transfer factor from power to oscillator motion is increased, which increases the open loop gain, and second, because a smaller relative power modulation on the transfer beam needs to be applied in order to compensate for the interferometer noise. Therefore it is interesting to have the highest transfer beam power as possible, which will ultimately be limited by the damage threshold of the micro-oscillator and the power availability.

The optimal value for the sensing beam mean power depends if the interferometer is quantum noise limited or if it is limited by technical radiation pressure noise. Let us calculate the sensing beam power $P_{\mathrm{s,q}}$ that minimizes the interferometer quantum noise contribution. This can be done by a compromise between the quantum radiation pressure noise and the readout noise contributions:

$$P_{\mathrm{s,q}} = \frac{c^2}{2\omega_0 \left|\chi_{m,0}\right|} \approx 0.4\,\mathrm{mW} \left(\frac{m}{40\,\mathrm{ng}} \cdot \frac{\Omega_0^2}{(100\,\mathrm{Hz})^2} \right) , \tag{3.62}$$

where $\left|\chi_{m,0}\right|$ is the susceptibility magnitude at low frequencies, which is approximately constant. Then, the out-of-loop power stability will be limited by the interferometer quantum noise if the following conditions are satisfied:

$$\mathrm{RPN}_{\mathrm{s}} < \frac{2\omega_0}{c} \sqrt{\hbar \left|\chi_{m,0}\right|} = 3 \times 10^{-8}\,\mathrm{Hz}^{-1/2} \left(\frac{(40\,\mathrm{ng})^{1/2}}{m^{1/2}} \cdot \frac{2\pi \cdot 100\,\mathrm{Hz}}{\Omega_0} \right) , \tag{3.63}$$

$$\frac{\mathrm{RPN}_{\mathrm{f}} \sqrt{P_{\mathrm{t}}}}{\left|\epsilon\right|} < 2 \sqrt{\frac{2\hbar\omega_0 \left|\chi_m\right|}{\left|\chi_{m,0}\right|}} = 1 \times 10^{-9}\,\mathrm{W}^{1/2} \cdot \mathrm{Hz}^{-1/2} \sqrt{\frac{\left|\chi_m\right|}{\left|\chi_{m,0}\right|}} \,. \tag{3.64}$$

The ASD of the out-of-loop relative power noise, calculated from Eq. 3.61, is plotted in Fig. 3.7 for two initial power noise values for the transfer and sensing beams. For the frequency range displayed in the figure, the high gain approximation is valid. The plot shows that a sub-shot noise stabilization can be achieved even when considering a high $\mathrm{RPN}_{\mathrm{s}} = 10^{-6}\,\mathrm{Hz}^{-1/2}$, that is easily achievable with a traditional pre-stabilization, and a realistic amplifier gain of $\left|\epsilon\right| = 10^3$. The red curve is limited by the interferometer technical radiation pressure noise. For this curve, P_{s} was optimized to be as low as possible, bounded by the next contribution coming from

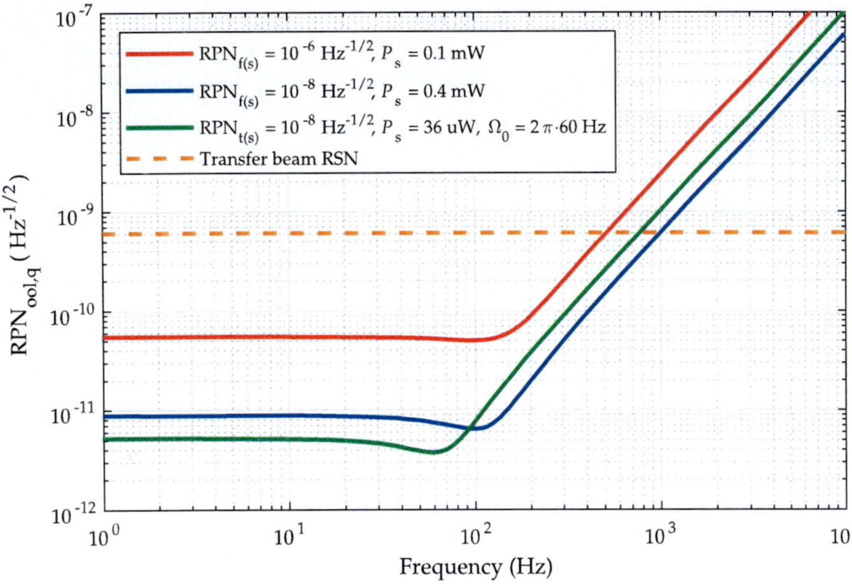

Fig. 3.7 ASD of the out-of-loop beam relative power noise for: $\mathrm{RPN_s} = \mathrm{RPN_f} = 10^{-6}\,\mathrm{Hz}^{-1/2}$ and $P_s = 0.1$ mW (red curve), and $\mathrm{RPN_s} = \mathrm{RPN_f} = 10^{-8}\,\mathrm{Hz}^{-1/2}$ and $P_s = 0.4$ mW (blue curve). The green curve has the same parameters of the blue curve, except for the oscillator fundamental longitudinal resonance of $\Omega_0 = 2\pi \cdot 60$ Hz and $P_s = 36\,\mu$W. For all curves $P_t = 1$ W and $|\epsilon| = 10^3$. As a reference, the transfer beam relative shot noise is depicted in the dashed yellow curve

the term $\mathrm{RPN_f}$. The blue curve, however, is limited by the interferometer quantum noise and P_s was adjusted according to Eq. 3.62. This limit can only be decreased further by increasing the mechanical susceptibility and reducing the sensing beam power simultaneously. This is depicted in the green curve, in which the longitudinal resonance of the oscillator was reduced to $\Omega_0 = 2\pi \cdot 60$ Hz, leading to an increase in $|\chi_{m,0}|$ by a factor of 2.8 (P_s was reduced accordingly). This increase in the power stability at low frequencies comes with a cost of decreasing the stabilization at high frequencies. Therefore the resonance frequency needs to be chosen according to the frequency range requirements of the experiment.

3.2 Thermal Noise Limit

An important noise source in precision measurements is thermal noise, which sets a fundamental limit in the degree to which the oscillator can stay at rest with the system in equilibrium at a certain temperature T. The physical process that generates these random fluctuations in the oscillator position originates from dissipations in the system.

The relation between the thermal fluctuations and the dissipation of energy in a system is given by the fluctuation-dissipation theorem [101–103]. The power spectral density of a system's fluctuating thermal motion x_{tn} is given by [11]:

$$x_{\mathrm{tn}}^2(\Omega) = \frac{4k_\mathrm{B}T}{\Omega^2}\mathrm{Re}\left[\frac{1}{Z(\Omega)}\right], \tag{3.65}$$

where Z is the mechanical impedance of the system and k_B is the Boltzmann constant. The impedance of a linear system can be calculated from the equation of motion that relates the external force F_{ext} necessary to cause the system to move with a sinusoidal velocity of amplitude v, i.e., $F_{\mathrm{ext}} = Zv$. By substituting $x = v/i\Omega$ into the equation of motion of a harmonic oscillator (derived in Sect. 2.1.3), the following impedance is calculated:

$$Z = \frac{-m\Omega^2 + k\left[1 + i\phi\right]}{i\Omega}, \tag{3.66}$$

where $k = m\Omega_0^2$ is the spring constant. Therefore, the power spectral density of the oscillator position is:

$$x_{\mathrm{tn}}^2(\Omega) = \frac{4k_B T k\phi}{\Omega\left[(k - m\Omega^2)^2 + k^2\phi^2\right]}. \tag{3.67}$$

The frequency behavior of the loss angle ϕ depends on the source of dissipation in the system, which can be loosely divided into external and internal losses.

The most common external loss source is due to viscous damping, which is caused by the surrounding gas molecules that hit the oscillator. This form of dissipation results in a restoring force proportional to the oscillator's velocity. The effect of viscous damping can be reduced by operating the experiment under low pressures. The viscous loss angle ϕ_v in this case can be obtained by the substitution $ik\phi_\mathrm{v}x \equiv b\dot{x}$ in the equation of motion of the harmonic oscillator, which leads to:

$$\phi_\mathrm{v}(\Omega) = \frac{\Omega}{\Omega_0 Q_\mathrm{v}}, \quad \text{with} \quad Q_\mathrm{v} = \frac{\Omega_0 m}{b}, \tag{3.68}$$

where Q_v is the viscous quality factor, and b is the damping parameter. Assuming that the experiment is operated at a pressure p small enough so that the mean free path for the residual gas molecules is large compared to the dimensions of the micro-oscillator, the damping parameter b is approximately [11]:

$$b = pA\sqrt{\frac{\mu}{2\pi k_B T}}, \tag{3.69}$$

which is proportional to the surface area A of the mirror and to the mass μ of one gas molecule. Here T is the gas temperature, which is related to the pressure by

$p = nk_B T$, with n the number density of molecules in the gas. If the residual gas is only composed by nitrogen, a high quality factor of:

$$Q_\mathrm{v} = 5 \times 10^7 \left(\frac{\Omega_0}{2\pi \cdot 100\,\mathrm{Hz}} \cdot \frac{m}{40\,\mathrm{ng}} \cdot \frac{10^{-6}\,\mathrm{mbar}}{p} \cdot \frac{\pi\, 35^2\,\mu\mathrm{m}^2}{A} \right), \tag{3.70}$$

can be obtained, for a gas at room temperature. Table-top experiments however, can be dominated by a monolayer of water that sticks strongly to the vacuum chamber walls and randomly de-adheres over time, which can impart a random kick of momentum in the micro-oscillator. In this case, inelastic scattering needs to be taken into account as the molecules can stick for a short time to the mirror surface. Considering that the mirror is a cylinder with radius R, surface area $A = \pi R^2$, and thickness ℓ, the following damping parameter can be calculated [104]:

$$b = pA\sqrt{\frac{128\mu}{\pi k_B T}} \left(1 + \frac{\ell}{2R} + \frac{\pi}{4}\right). \tag{3.71}$$

For the chosen micro-oscillator in this work, a thickness ℓ around $4.2\,\mu\mathrm{m}$ and a radius R of $35\,\mu\mathrm{m}$ will be considered. In this case, a quality factor of $Q_\mathrm{v} = 2 \times 10^6$ is expected, which is more than 1 order of magnitude lower than what calculated with Eq. 3.70.

The typical behavior of thermal noise displacement caused by viscous damping is shown by the red curve of Fig. 3.8. The displacement is approximately constant for frequencies below resonance:

$$x_\mathrm{tnv}(\Omega \ll \Omega_0) \approx \sqrt{\frac{4k_B T}{Q_\mathrm{v} m \Omega_0^3}}, \tag{3.72}$$

and it falls like $1/\Omega^2$ above resonance:

$$x_\mathrm{tnv}(\Omega \gg \Omega_0) \approx \frac{\Omega_0^2}{\Omega^2}\sqrt{\frac{4k_B T}{Q_\mathrm{v} m \Omega_0^3}}. \tag{3.73}$$

For off-resonance frequencies the displacement is reduced by increasing the quality factor. At resonance, however, the displacement is directly proportional to the quality factor and therefore it will be increased with a higher quality factor.

The most common form of internal loss is called structural damping, which is associated with internal friction losses. For this damping, the loss angle has only a weak dependence on frequency and is treated as a constant:

$$\phi_\mathrm{s} = \frac{1}{Q_\mathrm{s}}, \tag{3.74}$$

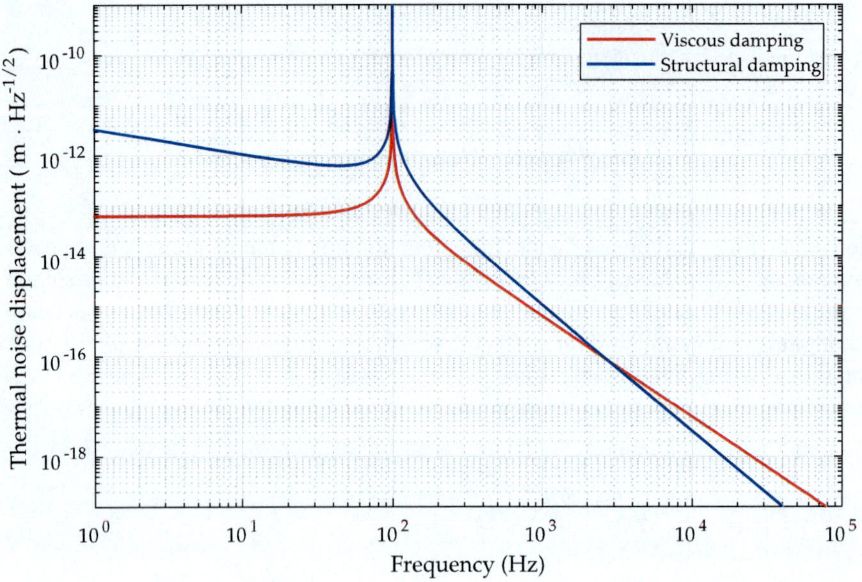

Fig. 3.8 Comparison of the ASD of the thermal noise displacement caused by viscous and structural damping. The calculation was done assuming a structural and viscous quality factor of $Q_s = 2 \times 10^5$ and $Q_v = 5 \times 10^7$, and an oscillator with $m = 40$ ng and $\Omega_0 = 2\pi \cdot 100$ Hz, operated at a temperature of $T = 4$ K

where Q_s is the structural quality factor. The typical thermal noise displacement modeled by structural damping is shown by the blue curve of Fig. 3.8. Unlike the viscous regime, the displacement is proportional to $\Omega^{-1/2}$ for frequencies below resonance,

$$x_{\text{tns}}(\Omega \ll \Omega_0) \approx \sqrt{\frac{4k_B T}{Q_s m \Omega_0^2} \cdot \frac{1}{\Omega}}\,, \tag{3.75}$$

and proportional to $\Omega^{-5/2}$ for frequencies above resonance,

$$x_{\text{tns}}(\Omega \gg \Omega_0) \approx \sqrt{\frac{4k_B T \Omega_0^2}{m Q_s} \cdot \frac{1}{\Omega^5}}\,. \tag{3.76}$$

The total thermal noise displacement from the viscous and structural damping contributions is calculated by substituting the total loss angle $\phi = \phi_v + \phi_s$, into Eq. 3.67.

Let us now calculate the minimum relative power noise $\text{RPN}_{\text{ool,tn}}$ achievable in the out-of-loop beam limited by thermal noise. Its value is equal to the power modulation that needs to be imprinted on the transfer beam to compensate the thermally driven motion of the oscillator (imposing $x_{\text{rp}} \overset{!}{=} x_{\text{tn}}$), which leads to:

$$\mathrm{RPN}_{\mathrm{ool,tnv}} = \frac{c}{P_{\mathrm{t}}}\sqrt{k_B T b}\,, \tag{3.77}$$

$$\mathrm{RPN}_{\mathrm{ool,tns}} = \frac{c\Omega_0}{P_{\mathrm{t}}}\sqrt{\frac{k_B T m}{Q_{\mathrm{s}}\Omega}}\,. \tag{3.78}$$

For the viscous regime, the stability is improved by lowering the area of the mirror. In the structural regime, the stability is improved by lowering the spring constant of the mirror, assuming that the structural quality factor is independent on it. A stability of $10^{-9}\,\mathrm{Hz}^{-1/2}$ at 10 Hz can be reached with the following values:

$$\mathrm{RPN}_{\mathrm{ool,tns}} = 10^{-9}\,\mathrm{Hz}^{-1/2}\cdot\left(\frac{\Omega_0}{2\pi\cdot 100\,\mathrm{Hz}}\cdot\frac{2.5\,\mathrm{W}}{P}\sqrt{\frac{T}{4\,\mathrm{K}}\cdot\frac{2\times 10^5}{Q_{\mathrm{s}}}\cdot\frac{m}{40\,\mathrm{ng}}\cdot\frac{2\pi\cdot 10\,\mathrm{Hz}}{\Omega}}\right). \tag{3.79}$$

The structural thermal noise contribution to $\mathrm{RPN}_{\mathrm{ool}}$ scales with frequency in the opposite way than quantum noise, as it is larger for low frequencies and smaller for high frequencies. The parameters with most impact on improving the out-of-loop power stability in this case are the transfer beam power and the mirror spring constant. The contribution from viscous damping to the $\mathrm{RPN}_{\mathrm{ool}}$ will eventually predominate at high frequencies since it is frequency independent. This is shown in Fig. 3.9, where the relative power noise is plotted separately for the structural and viscous contributions for two oscillators with different resonance frequencies. The dashed purple curve represents the contribution from the total thermal noise.

3.3 Total Fundamental Limit

The total fundamental limit of the $\mathrm{RPN}_{\mathrm{ool}}$ is obtained as an uncorrelated sum of the transfer scheme's quantum noise (Eq. 3.61) and thermal noise (Eq. 3.78) contributions.

Figure 3.10 shows the individual contributions of the interferometer noises to the out-of-loop beam stability. By comparing the quantum and thermal noise contributions, it is possible to observe that in general the $\mathrm{RPN}_{\mathrm{ool}}$ stability is limited by thermal noise at low frequencies and by quantum noise at high frequencies. For this reason, only structural thermal noise will have an impact on the total fundamental limit.[3] Since the interferometer readout noise dominates the quantum limit at high frequencies, it is interesting to increase the sensing beam power above the level defined by Eq. 3.62. This power increase will reduce the fundamental limit at high frequencies up to the point where the technical radiation pressure noise from the sensing beam starts to dominate the thermal noise. The optimal sensing beam

[3] Here it was assumed that the viscous thermal noise at low frequencies is negligible in comparison with the structural thermal noise.

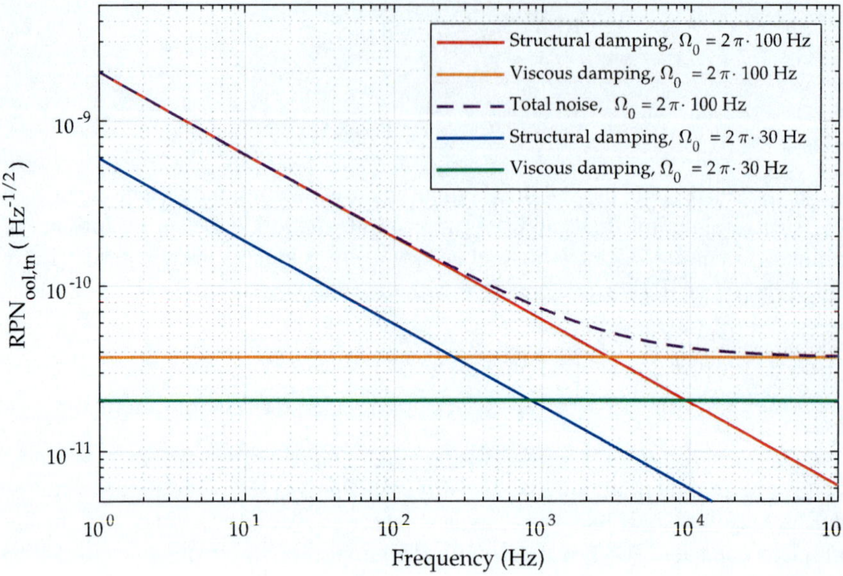

Fig. 3.9 ASD of the relative power noise $\mathrm{RPN}_{\mathrm{ool,tn}}$ of the out-of-loop beam limited by thermal noise in the micro-oscillator. The red, yellow, and purple curves were calculated for an oscillator with a resonance frequency of $\Omega_0 = 2\pi \cdot 100\,\mathrm{Hz}$, while the blue and green curves were calculated for $\Omega_0 = 2\pi \cdot 30\,\mathrm{Hz}$. The dashed purple curve represents the total thermal noise contribution. All curves were calculated with the following parameters: $Q_\mathrm{s} = 2 \times 10^5$, $Q_\mathrm{v} = 5 \times 10^7$, $m = 40\,\mathrm{ng}$, $P_\mathrm{t} = 4\,\mathrm{W}$, and $T = 4\,\mathrm{K}$

power is now obtained by imposing that the separated contributions from structural thermal noise, technical radiation pressure noise, and readout noise meet at a certain frequency. Thermal noise and technical radiation pressure noise curves meet at the following frequency:

$$\Omega_{\mathrm{tn,rp}} = \frac{4c^2\Omega_0^2 k_B T m}{Q_\mathrm{s}\,\mathrm{RPN}_\mathrm{s}^2 P_\mathrm{s}^2}, \tag{3.80}$$

while thermal noise and readout noise curves meet at:

$$\Omega_{\mathrm{tn,r}} = \left(\frac{8\Omega_0^2 k_B T \omega_0 P_\mathrm{s}}{Q_\mathrm{s} c^2 \hbar m}\right)^{1/5}. \tag{3.81}$$

Then, the optimal sensing beam power is obtained in the condition when $\Omega_{\mathrm{tn,r}} = \Omega_{\mathrm{tn,rp}}$. This leads to:

$$P_\mathrm{s} = \left(\frac{2^7 c^{12}\Omega_0^8 (k_B T)^4 m^6 \hbar}{Q_\mathrm{s}^4\,\mathrm{RPN}_\mathrm{s}^{10}\omega_0}\right)^{1/11} = 37\,\mathrm{mW}\left(\frac{10^{-8}\,\mathrm{Hz}^{-1/2}}{\mathrm{RPN}_\mathrm{s}}\right)^{10/11}, \tag{3.82}$$

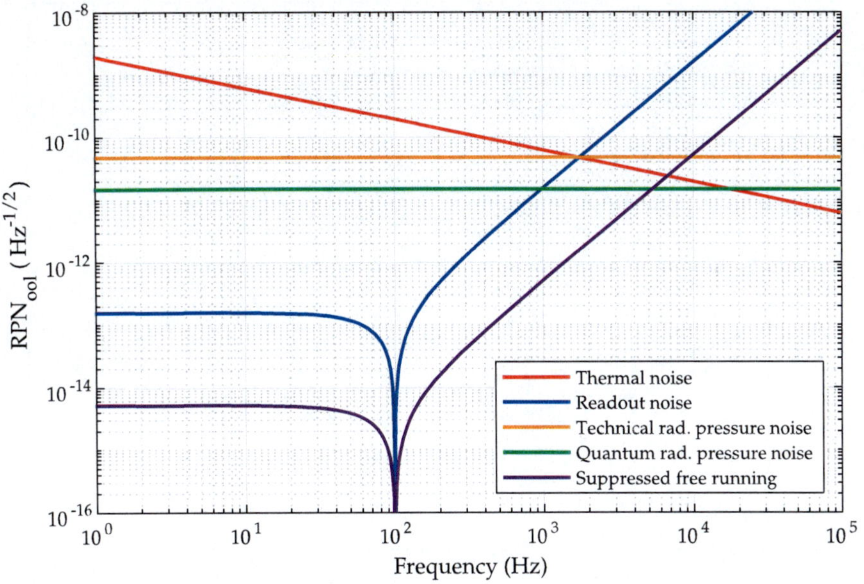

Fig. 3.10 ASD of the following contributions of interferometer noise projected to the out-of-loop relative power noise: structural thermal noise (red), interferometer readout noise (blue), technical radiation pressure noise (yellow), and quantum radiation pressure noise (green). The transfer beam free running power noise suppressed by the open loop gain is displayed by the purple curve. The following parameters were used: $Q_s = 2 \times 10^5$, $T = 4\,\mathrm{K}$, $P_t = 4\,\mathrm{W}$, $P_s = 37\,\mathrm{mW}$, $|\epsilon| = 10^3$ and $\mathrm{RPN_s} = \mathrm{RPN_f} = 10^{-8}\,\mathrm{Hz}^{-1/2}$

which is the power used in Fig. 3.10. As expected, the individual noise projections meet at 1.7 kHz. Quantum radiation pressure noise and the suppressed free running transfer beam power noise are a factor of 3 below the total noise and therefore do not limit the total stability. A higher sensing beam power can also be implemented in order to reduce the relative power noise at high frequencies, at the cost of increasing the noise at low frequencies.

The total fundamental limit for different oscillators is shown in Fig. 3.11. The red curve was plotted with the parameters for the micro-oscillator chosen for this chapter, and the blue curve for the same type of oscillator but with a lower resonance frequency of $\Omega_0 = 2\pi \cdot 60\,\mathrm{Hz}$. These parameters were chosen for an optimal stabilization aimed at frequencies higher than 10 Hz, which is the target in gravitational wave detectors. As a result, a power stability smaller than $6 \times 10^{-10}\,\mathrm{Hz}^{-1/2}$ can be achieved for frequencies between 10 Hz and 6 kHz. In addition to that, bright squeezed generation is possible to a level of 11 dB for the micro-oscillator analyzed in this chapter. For frequencies where the power stability is limited by structural thermal noise, the squeezing level can be calculated with the following equation:

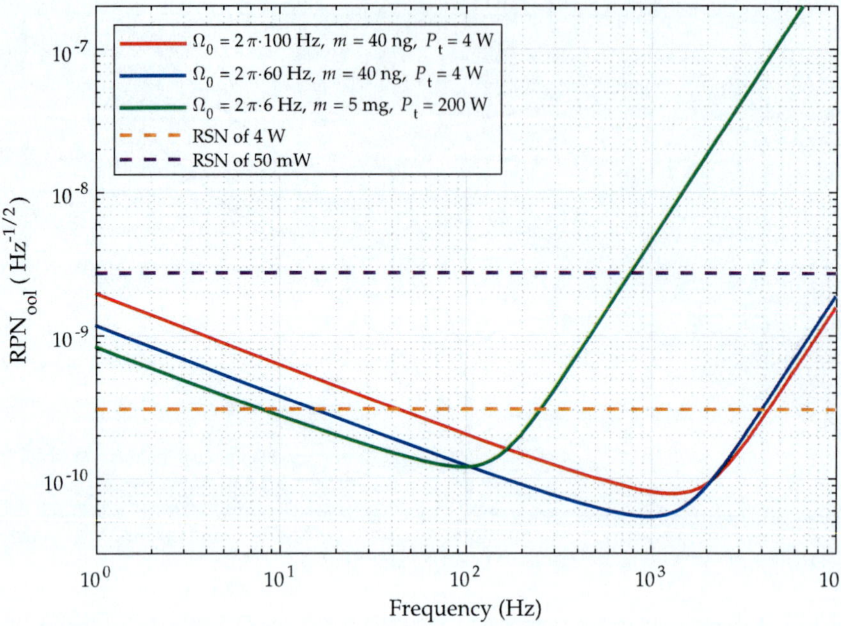

Fig. 3.11 ASD of the $\mathrm{RPN_{ool}}$ limited by the total interferometer noise for the following parameters: red curve: $m = 40\,\mathrm{ng}$, $\Omega_0 = 2\pi \cdot 100\,\mathrm{Hz}$, $P_t = 4\,\mathrm{W}$ and $P_s = 37\,\mathrm{mW}$, blue curve: $m = 40\,\mathrm{ng}$, $\Omega_0 = 2\pi \cdot 60\,\mathrm{Hz}$, $P_t = 4\,\mathrm{W}$ and $P_s = 26\,\mathrm{mW}$, and green curve: $m = 5\,\mathrm{mg}$, $\Omega_0 = 2\pi \cdot 6\,\mathrm{Hz}$, $P_t = 10\,\mathrm{kW}$, and $P_s = 2\,\mathrm{W}$. For all curves $\mathrm{RPN_f} = \mathrm{RPN_s} = 10^{-8}\,\mathrm{Hz}^{-1/2}$, $|\epsilon| = 10^3$, $T = 4\,\mathrm{K}$, and $Q_s = 2 \times 10^5$. For reference, the relative shot noise of a beam with 4 W is shown by the dashed yellow curve and the relative shot noise of a beam with 50 mW (traditional stabilization scheme stability) is shown by the dashed purple curve

$$\frac{\mathrm{RSN_{ool}}}{\mathrm{RPN_{ool}}} = \frac{1}{c\Omega_0}\sqrt{\frac{2\hbar\omega_0 P_t Q_s \Omega}{k_B T m}} \propto \sqrt{P_t}\,. \tag{3.83}$$

The fact that the amount of bright squeezing is proportional to the transfer beam power is an additional advantage of this scheme over the traditional scheme assisted by squeezing, in which the amount of squeezing is inversely proportional to the out-of-loop power (see Eq. 1.13) and therefore it disfavors the generation of a high power squeezed beam.

Micro-oscillators with similar individual parameters have been reported in references [70] (mass around 40 ng) and [105] (resonance frequencies around 150 Hz and quality factors of $Q = 2 \times 10^5$ at 10 K). Although a single device with the used parameters does not yet exist, the field of micro and nanofabrication have made advances that lead to unprecedented performances and such a device might be available in the near-future. An open and important question is the power damage threshold of the oscillators, which is a critical parameter for the stabilization level. For this reason, an additional curve was plotted (green curve) for a 5 mg suspended

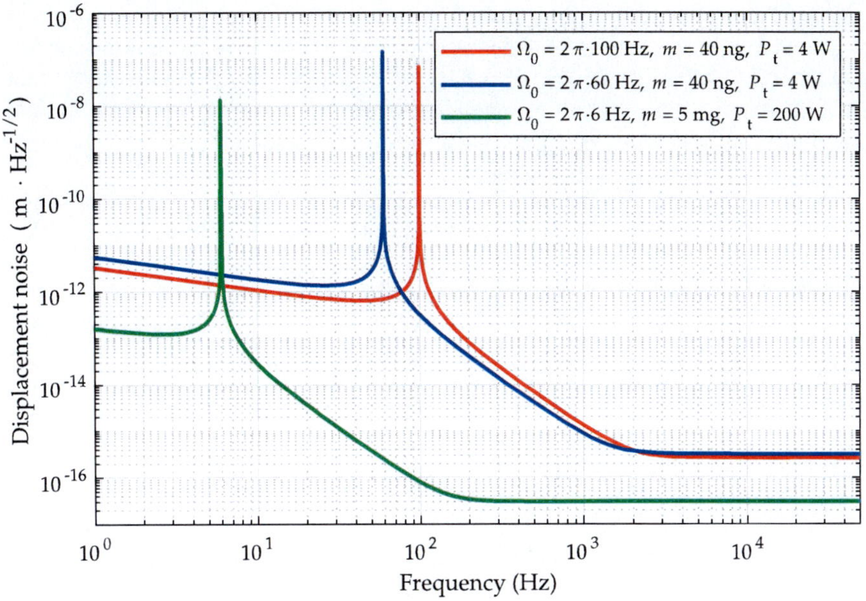

Fig. 3.12 ASD of the requirement for the technical displacement noise of the interferometer. The absolute incident power noise of the transfer beam was calculated using the RPN_{ool} obtained with the parameters of Fig. 3.11, and its corresponding mean powers

mirror with resonance frequency of $\Omega_0 = 2\pi \cdot 6$ Hz and a transfer beam power of 200 W. Similar parameters have been reported in the state of the art [69, 106]. Since the relative shot noise of such a high power beam is very low (4×10^{-11} Hz$^{-1/2}$), a stabilization below the relative shot noise is not possible with these parameters.

It is worth to note that for an alternative interferometer configuration in which the micro-oscillator is the end mirror for both interferometer arms ([107], mirror in the middle configuration), technical radiation pressure noise would not contribute to RPN_{ool}. In this case, P_s can be further increased, which would reduce the RPN_{ool} at high frequencies without compromising the stability performance at low frequencies.

Let us now analyze the interferometer sensitivity required to achieve the calculated RPN_{ool}. This calculation is important because in reality there are several technical noise sources in the interferometer such as vibrations, scattering, and electronic noise that needs to be below the noise sources accounted in this chapter. The requirement is calculated by imposing that the interferometer technical displacement noise is smaller than the radiation pressure displacement caused by the absolute power noise $RPN_{ool} \cdot P_t$. Figure 3.12 shows the expected displacement for the analyzed oscillators, calculated with the parameters of Fig. 3.11. Compact and table top interferometers can, with careful design, reach sensitivities down to 10^{-14}m · Hz$^{-1/2}$ up to kHz regime [46, 108, 109]. Sensitivities in the femtometer level are more challenging and therefore the use of high susceptibility resonators (red and blue curves) is favored with respect to the low susceptibility and high power configuration (green curve). Another aspect to consider is that experiments requiring a power stability below $10^{-9}\,\mathrm{Hz}^{-1/2}$ will most likely count on technologies and environments that are needed to suppress such technical noises. This is the case for gravitational wave detectors, which have suspended optical benches (vibration isolation), cryogenic temperatures (reduction of thermal noise), and low noise electronics.

3.4 Frequency Noise Imprinted in the Out-of-Loop Beam

The residual motion of the micro-oscillator when the power stabilization loop is turned on will imprint a phase/frequency noise on the out-of-loop beam. It is important to quantify this noise since the laser source in gravitational wave detectors is also required to be highly stable in frequency. The total imprinted frequency noise is calculated as an uncorrelated sum of the quantum and thermal noise contributions. The quantum noise in the phase quadrature of the out-of-loop beam when the power stabilization loop is turned on is given by Eq. 3.52. The ASD of the laser phase noise $\delta\phi$ can be then calculated by the following equation[4]:

$$\delta\phi = \sqrt{\frac{\hbar\omega_0 S_{\mathrm{ss}}}{2P_{\mathrm{t}}}}. \tag{3.84}$$

Therefore, by using the relation $\delta\omega = \delta\phi \cdot \Omega$, the additional frequency noise imprinted in the out-of-loop beam by the control loop can be obtained. The result is shown by the blue curve in Fig. 3.13, where the typical free running frequency noise of an non-planar ring oscillator (NPRO) laser (red curve) was used to calculate $S_{\mathrm{ss}}^{\hat{f}}$. The residual thermal noise displacement $x_{0,\mathrm{tn}}$ of the micro-oscillator will be equal to the thermal noise displacement x_{tn} when the loop is open, multiplied by the noise reduction factor of the control loop:

$$x_{0,\mathrm{tn}} = \frac{x_{\mathrm{tn}}}{\left|1 + \epsilon\sqrt{\mathcal{K}_\alpha \mathcal{K}_\beta}\right|}. \tag{3.85}$$

This is because the power stabilization loop will apply a power modulation on the beam transmitted by the power modulator in order to suppress the thermal noise fluctuations in the oscillator's position and keep the error signal zero. The frequency noise imprinted on the out-of-loop beam by this residual thermal noise motion is displayed by the green curve in Fig. 3.13. The plots show that the imprinted noise by the oscillator is several orders of magnitude below the free running noise of the NPRO at low frequencies, and one order of magnitude at high frequencies. Hence, the imprinted frequency noise should not be an obstacle to reach the requirements of

[4] This relation can be calculated from Eqs. 3.1 and 3.5.

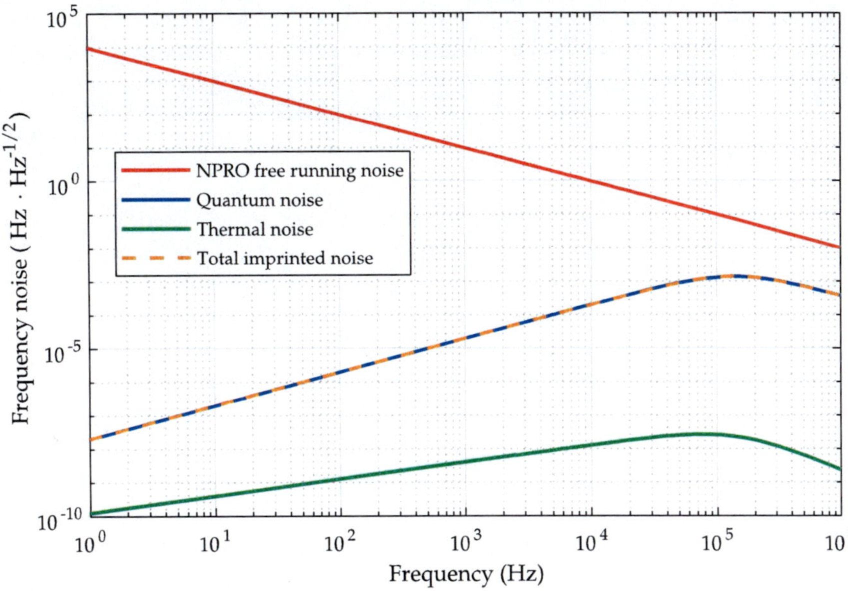

Fig. 3.13 ASD of the frequency noise imprinted on the out-of-loop beam by the micro-oscillator's residual motion due to contributions from quantum noise (blue curve), and thermal noise (green curve). The typical free running frequency noise of an NPRO laser is shown by the red curve

the laser source, if the laser frequency stabilization in the gravitational wave detector is performed in stages located after the proposed power stabilization scheme.

The fact that the motion of the micro-oscillator is stabilized by the power stabilization control loop is an additional advantage of using a radiation pressure phase transfer scheme, since the contributions from nonlinear effects caused by a large mirror motion should be negligible even for high laser powers. This is not the case for nonlinear materials, where effects like stimulated Brillouin scattering and thermal fluctuations have been reported for high laser powers, which is required in order to compensate for the low transfer phase coefficient.

3.5 Comments on Ponderomotive Squeezing

It is important to note that the generation of amplitude squeezing via the negative feedback in the proposed scheme does not violate Weisman and Milburn condition [110] which states that "feedback mediated by homodyne detection can only produce nonclassical light, if the system dynamics can do so without feedback", since optomechanical systems are known to produce ponderomotive squeezing on their own [69].

Let us now calculate the natural ponderomotive squeezing factor in reflection of the micro-oscillator. In the absence of the sensing beam, the out-of-loop amplitude and phase quadratures are given by:

$$\hat{o}_{\mathrm{c}} = \hat{f}_{\mathrm{c}}\,, \tag{3.86}$$

$$\hat{o}_{\mathrm{s}} = \hat{f}_{\mathrm{s}} - \mathcal{K}_\beta \cdot \hat{f}_{\mathrm{c}}\,, \tag{3.87}$$

which are not squeezed. However, squeezing happens in an intermediary quadrature $\hat{o}_{\mathrm{z}}$ given by:

$$\begin{aligned}\hat{o}_{\mathrm{z}} &= \hat{o}_{\mathrm{c}} \cdot \cos\theta_{\mathrm{z}} + \hat{o}_{\mathrm{s}} \cdot \sin\theta_{\mathrm{z}} \\ &= (\cos\theta_{\mathrm{z}} - \mathcal{K}_\beta \sin\theta_{\mathrm{z}}) \cdot \hat{f}_{\mathrm{c}} + \sin\theta_{\mathrm{z}} \cdot \hat{f}_{\mathrm{s}}\,.\end{aligned} \tag{3.88}$$

This quadrature has the following power spectral density:

$$S_{\mathrm{zz}}^{\hat{o}} = \left(\cos^2\theta_{\mathrm{z}} + 2\mathrm{Re}\left(\mathcal{K}_\beta\right)\cos\theta_{\mathrm{z}}\sin\theta_{\mathrm{z}} + \left|\mathcal{K}_\beta\right|^2 \sin^2\theta_{\mathrm{z}}\right) \cdot S_{\mathrm{cc}}^{\hat{f}} + \sin^2\theta_{\mathrm{z}} \cdot S_{\mathrm{ss}}^{\hat{f}}\,, \tag{3.89}$$

with minimum and maximum values of:

$$S_{\mathrm{zz}}^{\hat{o}} = \frac{S_{\mathrm{cc}}^{\hat{f}}}{2}\left(1 + \left|\mathcal{K}_\beta\right|^2 + \frac{S_{\mathrm{ss}}^{\hat{f}}}{S_{\mathrm{cc}}^{\hat{f}}} \pm \sqrt{\left(1 + \left|\mathcal{K}_\beta\right|^2 + \frac{S_{\mathrm{ss}}^{\hat{f}}}{S_{\mathrm{cc}}^{\hat{f}}}\right)^2 - 4 \cdot \frac{S_{\mathrm{ss}}^{\hat{f}}}{S_{\mathrm{cc}}^{\hat{f}}}}\right)\,. \tag{3.90}$$

For a transfer beam initially in a coherent state, the squeezed quadrature will have the following power spectral density:

$$S_{\mathrm{zz}}^{\hat{o}} = \left(\sqrt{1 + \left(\left|\mathcal{K}_\beta\right|/2\right)^2} - \left|\mathcal{K}_\beta\right|/2\right)^2 \approx \frac{1}{\left|\mathcal{K}_\beta\right|^2} \quad \text{for} \quad \left|\mathcal{K}_\beta\right| \gg 1\,, \tag{3.91}$$

with $\theta_{\mathrm{z}} = \arctan\left(-2/\left|\mathcal{K}_\beta\right|\right)/2$. Hence, as expected from the Weisman and Milburn condition, the amplitude squeezing factor achieved via the feedback control loop cannot surpass the squeezing factor in the intermediary quadrature $\hat{o}_{\mathrm{z}}$, generated by the natural ponderomotive squeezing of the micro-oscillator (for a coherent state!). However, to obtain squeezing in the amplitude quadrature without the feedback, the quadrature $\hat{o}_{\mathrm{z}}$ needs to be rotated such as to be aligned to the laser carrier. A frequency dependent rotation could be achieved, for example, by using an optical cavity [111]. In this case, the cavity will also introduce its own quantum noise and the squeezing factor will in general be smaller than the value calculated in Eq. 3.91.

An important difference between the ponderomotive squeezing over the squeezing generated via the negative feedback in the proposed scheme arises when considering realistic laser power and phase noise in the transfer beam. By substituting a value of a free running relative power noise of $10^{-6}\,\mathrm{Hz}^{-1/2}$, and a mean power of 4 W in

Eq. 3.19, a value of $S_{cc} = 10^7$ is obtained. Similarly, by substituting a typical free running phase noise of an NPRO of $10^{-2}\,\text{rad} \cdot \text{Hz}^{-1/2}$ at 1 kHz in Eq. 3.84, a value of $S_{ss} = 4 \times 10^{15}$ is obtained, which is much larger than S_{cc}. In this case,

$$S_{zz}^{\hat{o}} \approx S_{cc}^{\hat{f}}, \tag{3.92}$$

meaning that the ponderomotive squeezing cannot surpass the free running power fluctuations of the laser. Hence, when considering classical laser noise, a higher squeezing factor will be obtained in the out-of-loop beam by implementing the feedback scheme proposed in this chapter than what would be obtained from the natural ponderomotive squeezing of the micro-oscillator.

3.6 Summary

This chapter presented the fundamental limits of a power stabilization via a radiation pressure transfer scheme. The in-loop power sensor of this scheme consisted of a Michelson interferometer with a micro-oscillator mirror in one of its arms, whose position is modulated according to the power modulations of a strong transfer beam. The interferometer readout was chosen such as is only sensitive to the phase quadrature out its output field via a balanced homodyne detection. Noise sources coupling at the interferometer readout represent the sensing noise of this scheme and will limit the maximum power stability achievable in the out-of-loop beam. This maximum stability was calculated by considering an interferometer fundamentally limited by quantum noise and by thermal noise of a micro-oscillator mirror with parameters reported in the state of art. The calculations showed that this is a realistic approach when considering a relative power noise of $10^{-8}\,\text{Hz}^{-1/2}$ in the interferometer beam, and that the interferometer has a displacement sensitivity higher than $10^{-14}\,\text{m} \cdot \text{Hz}^{-1/2}$ at frequencies below 500 Hz, of $10^{-15}\,\text{m} \cdot \text{Hz}^{-1/2}$ at frequencies around 1 kHz, and of $2 \times 10^{-16}\,\text{m} \cdot \text{Hz}^{-1/2}$ at frequencies higher than 2 kHz.

The analysis showed a remarkable difference in the out-of-loop stability performance in comparison to the traditional and OAC-coupling schemes since a sub-shot noise power stabilization can be achieved. This is a consequence from performing the power sensing via a non-demolition measurement. Under realistic experimental parameters, a strong bright squeezed beam with power of 4 W and up to 11 dB of squeezing might be achievable in the near future. Furthermore, a higher squeezing factor can be achieved by increasing the transfer beam power, since it was demonstrated that this factor is proportional to $P_t^{1/2}$. This is a considerable advantage over the traditional scheme assisted by squeezing in which the squeezing factor is reduced by increasing the out-of-loop beam power, and would require 30 dB of amplitude squeezed vacuum injected in the open port of the beamsplitter (for 50 mW in the in-loop detector). The calculations also showed that an $\text{RPN}_{\text{ool}} < 6 \times 10^{-10}\,\text{Hz}^{-1/2}$ can be achieved for frequencies between 10 Hz and 6 kHz. Such value could only be achieved in the traditional scheme by detecting 1 W in the in-loop detector. The out-of-loop power noise and the requirements in the interferometer sensitivity can be reduced by increasing the micro-oscillator susceptibility and the mean power in the transfer beam. In conclusion, the results of this chapter showed that the proposed scheme has a strong potential towards achieving a high power stability that could fulfill the requirements of future gravitational wave detectors. In addition to that, the implementation of this scheme can also be beneficial for high precision metrology experiments, especially optomechanical experiments.

Chapter 4
Power Stabilization via Radiation Pressure—Experimental Setup

Abstract This chapter describes the design of a proof-of-principle experiment of laser power stabilization via a radiation pressure response on a movable mirror. This scheme was chosen based on the theoretical investigations performed in Chaps. 2 and 3. The investigations showed that a high signal-to-noise ratio can be achieved in the in-loop sensor and, as a result, a power stability below the shot noise of the out-of-loop beam should be possible. The main goals of this experiment are to investigate the transfer from power fluctuations to mirror motion, to analyze technical noise sources and limitations in the readout of the mirror position, and to demonstrate a power stabilization by means of this novel sensing scheme. The chapter starts with a description of the chosen oscillator, and in sequence the details of the experimental setup are presented.

4.1 Choice of the Movable Mirror

Two of the most important parameters when choosing the oscillator mirror is its mass and its fundamental longitudinal resonance frequency, since they determine the magnitude of the oscillator displacement at low frequencies. Oscillators with low mass and low resonance frequency are desirable since they increase the radiation pressure transfer coefficient τ and decrease the requirements on readout technical noises, as discussed in Sects. 2.1.3 and 3.3. In addition to that, it is also important that the mirror has a high reflectivity (to maximize the momentum transfer), low absorption (for a high laser damage threshold), and low mechanical dissipation (for low thermal noise, see Chap. 3). For those reasons, the oscillator chosen for this proof-of principle experiment is a micro-oscillator consisting of a circular mirror pad suspended by a cantilever structure, as shown in Fig. 4.1a.

The mirror pads of a collection of micro-oscillators available for this experiment are made of 23 alternating layers of GaAs (high refraction index) and $Al_{0.92}Ga_{0.08}As$

M. Trad Nery, *Laser Power Stabilization via Radiation Pressure*, Springer Theses,
https://doi.org/10.1007/978-3-030-95868-8_4

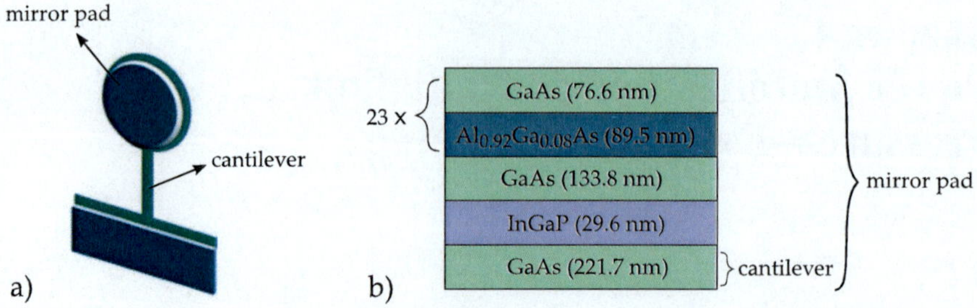

Fig. 4.1 Illustration of the micro-oscillator design: **a** mechanical model of the oscillator comprised of a circular mirror pad suspended by a cantilever structure, **b** cross sectional schematic of the epitaxial multilayer consisting of a GaAs structural support layer, a InGaP etch stop layer (for fabrication purposes), a GaAs layer, and 23 alternating layers of $Al_{0.92}Ga_{0.08}As$ and GaAs (Bragg mirror). The green layer in the mechanical model (**a**) corresponds solely to the 221.7 nm thick GaAs substrate and the blue layer represents the Bragg mirror containing all the layers displayed on the cross sectional schematic. Illustration adapted from [70]

(low refraction index), forming a distributed Bragg reflector (Fig. 4.1b), with a total thickness of 4.2 μm. Defined by the number of layers, the mirror pad has a transmission of around 250 ppm for a wavelength of 1064 nm. The number of layers was chosen as a compromise between high reflectivity and low mirror mass. The mirror pads are solely suspended by a 221.7 nm thick film of GaAs and therefore the optical and mechanical properties of the oscillator are decoupled. This design, together with a reduced number of Bragg layers,[1] allowed the micro-oscillators to have low masses down to 40 ng. The Bragg layers are deposited on top of the GaAs substrate via molecular beam epitaxy and the oscillator structure is defined using a double etch stop technique. Detailed informations on the fabrication process can be found in [70, 73]

A wafer containing several micro-oscillators was mounted on an 1 inch copper chip as shown in Fig. 4.2a.[2] The chip contained more than 100 micro-oscillators with different geometries following the design of Fig. 4.2b. The radius of the mirror pads varied from 25 μm to 100 μm, which resulted in a mass ranging from 40 ng to 200 ng. The length and width of the cantilevers varied from 19 μm to 285 μm and 4 μm to 30 μm, respectively. The spring constant of the oscillator decreases by decreasing the width of the cantilever, and by increasing the cantilever length. The chip used contained micro-oscillators with longitudinal resonances ranging from 117 Hz to approximately 1 kHz. Additionally, the oscillators were expected to have a mechanical quality factor for the longitudinal resonance on the order of 10^4 at room temperature.

[1] Previous generations of micro-oscillators had a Bragg mirror with 36 layers and a transmission of 10 ppm [70].

[2] The picture was taken from a chip containing another generation of micro-oscillators, used in an earlier stage of this experiment, and therefore with a different layout than what displayed in Fig. 4.2b.

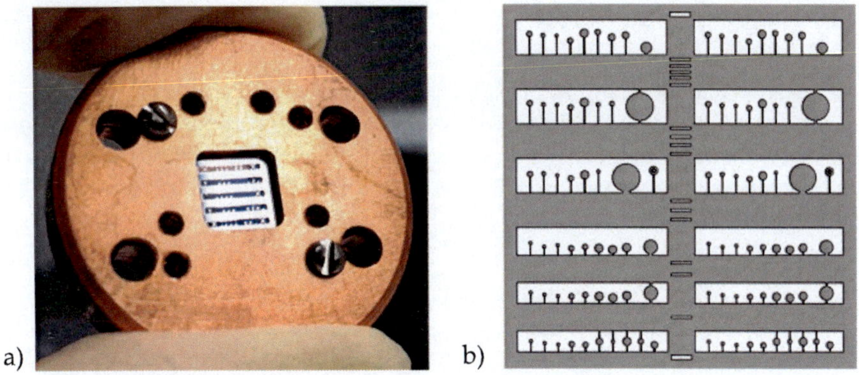

Fig. 4.2 **a** photograph of an 1 inch diameter copper mount containing the micro-oscillator wafer, and **b** model of the array of micro-oscillators with different geometries for the wafer used in this thesis (layout taken from reference [73])

Figure 4.3 shows a picture of three micro-oscillators present in the chip. The mirror pads with radius smaller than 25 μm were imperfectly etched and therefore the Bragg structure can be clearly seen in the picture. The oscillators were designed by the group of Thomas Corbitt in Louisiana State University (LSU, USA) in collaboration with the MIT, and were fabricated by the company Crystalline Mirror Solutions (CMS, USA).

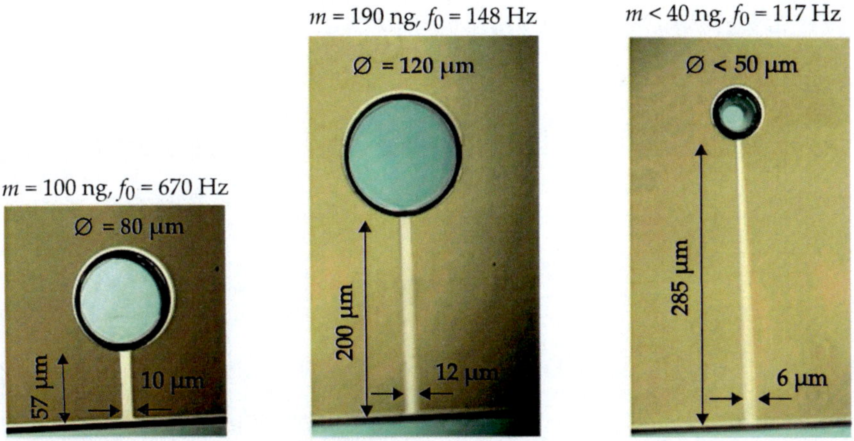

Fig. 4.3 Photograph of three micro-oscillators taken with an electronic microscope. The length and width of the cantilevers, the mirror pad diameter ∅, and the mass m and longitudinal resonance frequency f_0 of the oscillators are annotated

4.2 Experimental Setup

The experimental setup follows the general scheme presented in Sect. 3.1.3 and is shown in detail in Fig. 4.4. The setup was divided in two parts: a laser preparation area and an in-vacuum breadboard containing the Michelson interferometer, which was the core part of the experiment. The details of the setup are presented in the following subsections.

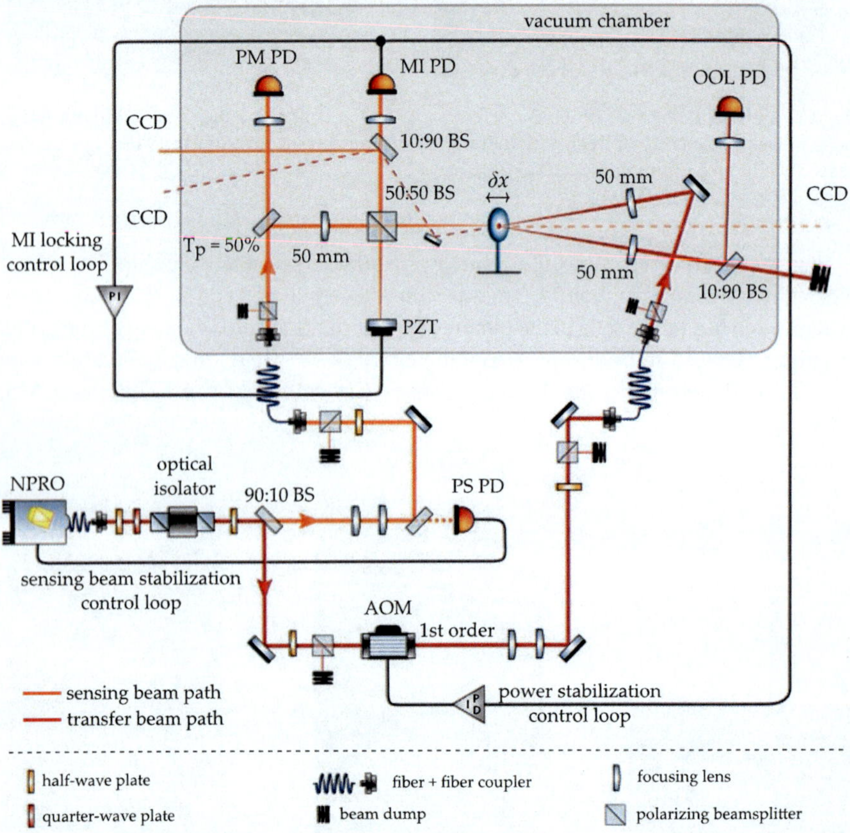

Fig. 4.4 Schematic of the power stabilization via radiation pressure experimental setup. An NPRO laser was split into the transfer beam (red trace) and sensing beam (orange trace) by a 90:10 beamsplitter. Both beams are guided by optical fibers to the vacuum chamber, where a breadboard containing the Michelson interferometer is located. The MI PD was the in-loop sensor for two control loops: the Michelson interferometer (MI) loop, which used the PZT as an actuator, and the power stabilization loop, which used the AOM as an actuator

4.2.1 Laser Preparation

The first component in the laser preparation area is the laser source: a continuous-wave and single-frequency solid-state Nd:YAG nonplanar ring oscillator (NPRO). This laser operated at a wavelength of 1064 nm with a fiber-coupled output and it delivered a maximum output power of 1 W. An optical isolator was placed after the NPRO to prevent back reflected light into the laser, which could disturb the stable operation of the laser. After the isolator, a beamsplitter reflected approximately 90% of the light to the transfer beam path (red trace) and transmitted the remaining light to the sensing beam path (orange trace).

The transfer beam was sent to an acousto-optic modulator (AOM) which was the power actuator in the power stabilization via radiation pressure control loop. The AOM consists of a crystal attached to a piezoelectric transducer (PZT), as shown in Fig. 4.5. When a high voltage sinusoidal signal with a radio frequency f_{RF} is applied to the PZT, a traveling acoustic wave propagates from the PZT to an absorber, creating areas of compression and rarefaction inside the crystal which have different refractive indexes. These areas act like a diffraction grating for the incoming light, and therefore the amplitude of the output light in a certain diffraction order can be controlled by adjusting the amplitude of the sound wave. As the light is scattered by a moving grating, the diffracted beam of order n will acquire an optical frequency shift of $\omega_0 \rightarrow \omega_0 \pm n \cdot 2\pi f_{\mathrm{RF}}$. This frequency shift has a great utility in this setup, since it prevents scattered light from the transfer beam from coupling to the interferometer displacement noise. This was the main reason an AOM was chosen as the power actuator in this experiment. The transmitted first order beam was used with 84.5% of diffraction efficiency, and the AOM driver had a radio frequency of 80 MHz.

The AOM crystal used in this setup was lead molybdate ($PbMoO_4$) from the company ISOMET. The crystal is birefringent and if the input light polarization is not aligned to one of the crystal axis, the output light suffers a polarization rotation

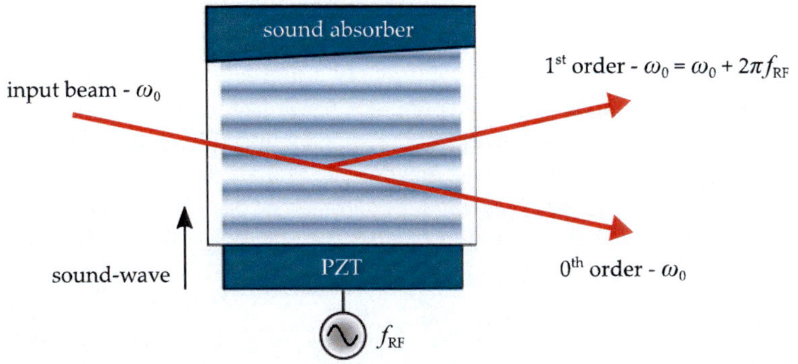

Fig. 4.5 Working principle schematics of an AOM: an input beam with an optical frequency ω_0 is diffracted by a moving diffraction grating which is generated by a sound wave inside of the AOM crystal

that can fluctuate due to temperature changes in the crystal, for example. This effect was observed as a power drift after a polarizing beamsplitter (PBS), over periods of one minute time scales and with large amplitude (around 10% of the total power). This posed a problem for the setup since the variations of the laser mean power changed the mean displacement of the oscillator and altered the contrast of the interferometer and the alignment of the oscillator with respect to the transfer and sensing beams. To minimize this issue, a PBS was placed before the AOM as a polarization filter and its output polarization was aligned with the crystal horizontal axis. Nevertheless, for a future setup, a crystal with smaller birefringence such as quartz would be a wiser choice.

The transfer and sensing beams were guided into the vacuum chamber via two single-mode and polarization maintaining type PANDA optical fibers, model 980 from the company Schäfter und Kirchhoff. Both fibers have FC-APC connectors on their ends. Mode matching lenses and alignment mirrors were placed in both paths for the light coupling into the fiber. In addition to that, PBSs on rotation mounts were placed before the fibers such that the incoming polarization is aligned with the fiber slow axis. The PBSs, together with half-wave plates, were used as power attenuation stages. The choice of fibers as a guiding medium for the light was made in order to avoid effects caused by beam jitter between the breadboard and the optical table and also to allow full mobility of the breadboard during early stages of the experiment. With this configuration, no misalignment in the oscillators or in the interferometer was observed when moving the breadboard outside and inside the chamber.

The laser preparation area also included a traditional power stabilization scheme for the sensing beam which will be discussed in Sect. 4.2.5.

4.2.2 *In-vacuum Breadboard*

The core part of the experiment containing the Michelson interferometer sat on top of an aluminum breadboard with dimensions of $27 \times 27 \times 1.2\ \mathrm{cm}^3$ placed inside the vacuum chamber, as shown in Fig. 4.6.

A collimated sensing beam of approximately 1 mm radius exited the fiber collimator (bottom of the picture) and was directed towards a 50:50 plate beamsplitter. The transmitted light was detected by the power monitor photodetector (PM PD), which measured the power noise at the input of the interferometer, while the reflected light was directed to a 50 mm focusing lens and then to the interferometer.

The interferometer was comprised of a 50:50 cubic beamsplitter (BS), a 0.5 inch high reflective mirror, and the micro-oscillator. The 0.5 inch mirror was glued to a PZT element which was used to scan the interferometer arm length, and also as an actuator to lock the interferometer at a specific operational point. A cubic beamsplitter was chosen since a plate beamsplitter would cause significantly different astigmatisms in the output beams due to the high divergence of the incoming beam (necessary to get a target beam waist on the micro-oscillator of 15 μm). This was because the directly reflected beam passed through the substrate only once while the transmitted

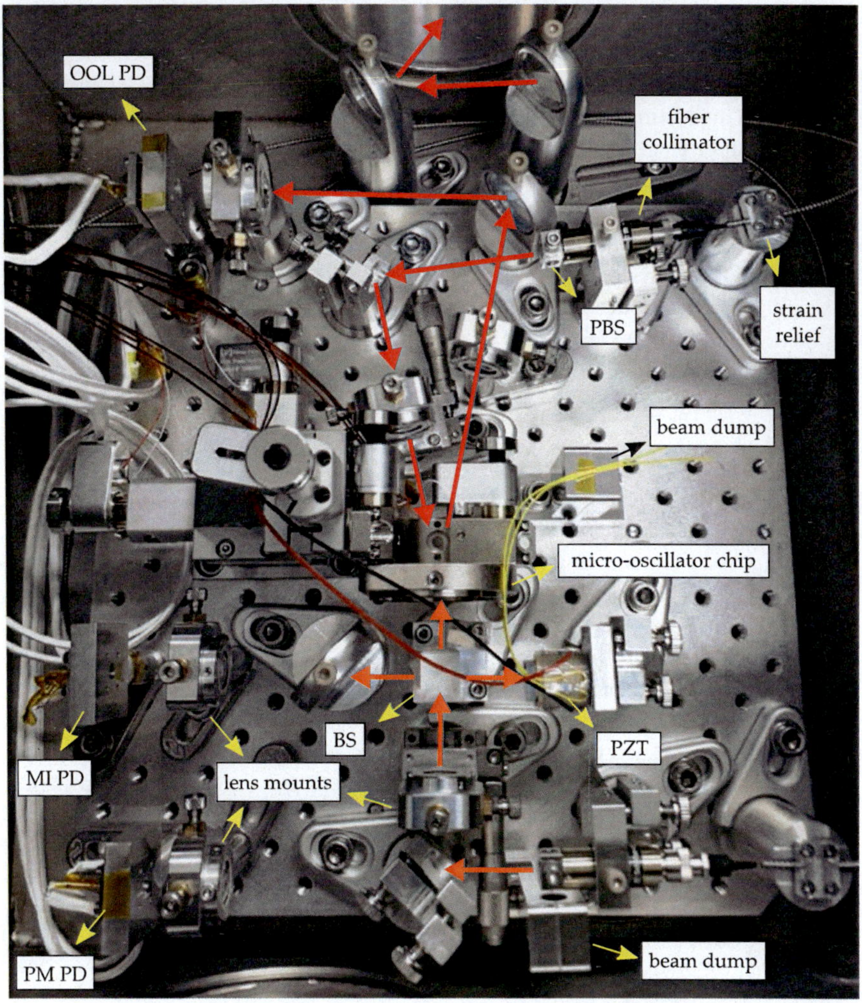

Fig. 4.6 Picture of the breadboard placed inside the vacuum chamber. The sensing beam path through the interferometer is marked by the orange arrows while the transfer beam path is marked by the red arrows

beam passed through three times before exiting the interferometer. A difference in the astigmatism limits the interferometer contrast, which for a plate beamsplitter was below 50%.

At the output of the interferometer, a beamsplitter transmitted 90% of the light to the Michelson interferometer photodetector (MI PD) and reflected the remaining light to a charge-coupled device (CCD) camera, placed outside the chamber, that monitored the interference pattern during the interferometer alignment. The transfer beam exited its fiber collimator (top of the picture) also with approximately 1 mm beam radius, passed through a 50 mm focusing lens and was directed to the micro-oscillator. The reflected beam went again through a 50 mm lens, which was placed to minimize the beam divergence, and in sequence was directed to a beamsplitter that reflected approximately 10% of the light towards the out-of-loop photodetector (OOL PD), and transmitted the remaining light to outside the vacuum chamber into a beam dump.

The schematics of the experimental setup also shows two CCD cameras, placed outside the vacuum chamber, that are positioned to capture the sensing and transfer beam light in transmission of the micro-oscillators (for alignment purposes).

A small PBS was placed at the output of both fiber collimators in order to filter polarization fluctuations induced by birefringences in the fiber. This was crucial for the sensing beam path to guarantee the coherence between the PM PD and the MI PD that would be otherwise reduced due to the polarization dependence of the beamsplitters.

The mirror mounts used in the breadboard are all from the company Radiant Dyes and have 4 springs for lower settling drifts. To allow control of the five crucial degrees of freedom of the micro-oscillators from outside the vacuum chamber, the 1 inch copper chip was mounted on a motorized mirror mount (pitch and yaw adjustment) attached to a XYZ translation stage (with a step size of less than 30 nm) from the company Newport, shown in Fig. 4.7a.

All photodiodes were rigidly placed on top of the breadboard (mount shown in Fig. 4.7b), in order to reduce beam jitter between the breadboard and the photodiodes. In addition to that, the photodiodes were placed as close as possible to the micro-oscillator chip to reduce the beam jitter at the oscillator's resonance frequencies. Due to the vacuum incompatibility, the photodetector electronics (transimpedance

a) 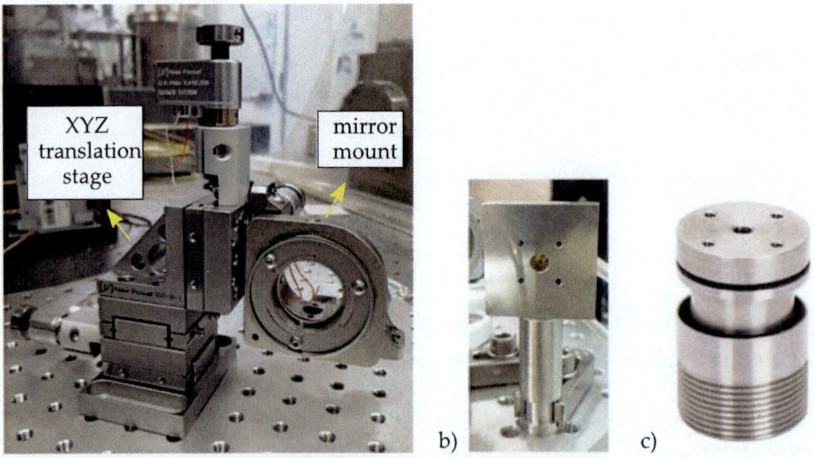 b) c)

Fig. 4.7 Pictures of breadboard components: **a** motorized mirror mount attached to the XYZ motorized translation stage, **b** vacuum photodiode mount, and **c** VIB100 breadboard isolation foot, displayed upside-down (picture taken from [112]) amplifier) were placed outside the vacuum chamber. This configuration is more prone to pick-up noise since the long cables are susceptible to electro-magnetic interference and therefore the cable lengths were kept shorter than 1 m. The long cables also add additional capacitance that needs to be compensated, for stability purposes, by the feedback capacitance of the transimpedance amplifier [113]. This compensation reduces the bandwidth of the photodetector, which for this experiment was around 1 MHz. All photodiodes were from JENOPTIK Polymer Systemsmade and are made of InGaAs. Both the PM PD and MI PD had an active diameter of 1 mm, while the OOL PD had a bigger diameter of 3 mm to avoid possible clipping effects since it was placed further away from the oscillator. Figure 4.8 shows the electronic noise of the photodetectors measured by blocking the light at the photodiodes (dark noise). For comparison, the electronic noise of the Stanford Research Systems spectrum analyzer (used in most measurements displayed in this thesis) when terminated with a 50 Ω resistor is shown. Apart from the peaks at 10 kHz and 50 kHz, the PDs dark noise in the measured bandwidth are on the same level as expected if the photodiodes were directly connected to the transimpedance amplifier.

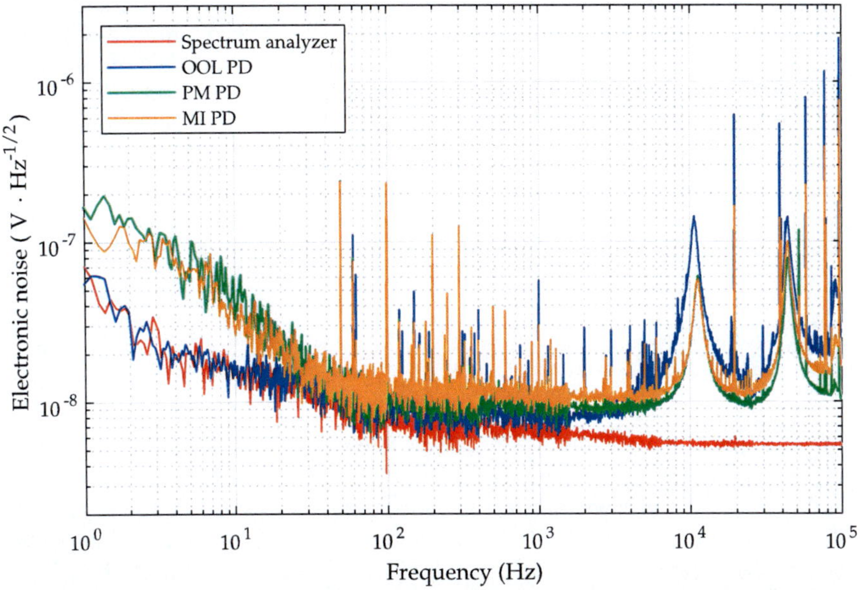

Fig. 4.8 ASD of the electronic noise of: Stanford Research Systems SR785 spectrum analyzer terminated with a 50 Ω resistor (red curve), out-of-loop photodetector (blue curve), power monitor photodetector (green curve) and Michelson interferometer photodetector (orange curve). The electronic noise of the photodetectors was measured by blocking the light at the photodiodes (dark noise)

4.2.3 Vacuum System

The vacuum system was comprised of a vacuum chamber, a gate valve, a turbo pump, a magnetic valve, and a scroll pump, as shown in the picture displayed in Fig. 4.9. The chamber had dimensions of $40 \times 40 \times 20$ cm^3, was made of stainless steel and contained a top lid, from where the breadboard was inserted, and 4 flanges. The flanges were used to connect two optical viewports for light exit, one electronics and fiber feed-through, and a gate valve.

A picture of the in-house made feed-through is shown in Fig. 4.10a. It contains two sub-D connectors used for the electronics of the PDs, the translation stage and the PZT, and two fiber ports made of a polytetrafluorethylen (PTFE) ferrule (Fig. 4.10b) enclosed by a nut connector. Because of the FC-APC connectors, the bare fibers were first cut in half, passed through the port and then spliced back together. After the splicing, the nut connector was tightened until the PTFE, which is a high deformable material, was hermetically compressed around the fiber, ensuring the air sealing of the connection. Apart from a small piece close to the fiber connection, the bare fibers were surrounded by aluminum tubes for protection, and a strain relief mount was placed between the tubes and fiber collimators (see Fig. 4.6).

The scroll pump provided a back pressure of 10^{-3} mbar for the turbo pump, which could ultimately reach pressures of 10^{-7} mbar. Since both pumps introduce significant amount of vibrations, all the measurements were done with the pumps turned off, and the low pressure inside of the chamber was maintained by closing the gate valve at the input of the chamber. An automatic magnetic valve connected to a pressure sensor was placed between the scroll and turbo pump to protect the system and the optics in case of a scroll pump failure or power outage, which would have caused dust particles from the scroll pump to flow into the chamber.

4.2.4 Vibration Isolation

Vibrations such as seismic motion or acoustic noise can couple into the experimental setup by different paths and introduce undesired displacement noise in the interferometer. At the laser preparation area, vibrations can, for example, cause beam jitter at the input of the fibers which, due to fluctuations in the light coupling, are converted to power fluctuations at the fiber outputs. At the breadboard, vibrations can couple directly to the interferometer by changing the differential arm length, or indirectly via interference with scattered light for example (details are discussed in Sect. 5.1.4).

To suppress the influence of vibrational noise, the experimental setup comprised of the laser preparation and the vacuum chamber was mounted on top of an optical table suspended by four pneumatic feet (see Fig. 4.9). The feet provided vertical and horizontal isolation for frequencies above 4 Hz. To reduce the coupling at the interferometer even further, the breadboard was mounted on top of three compact

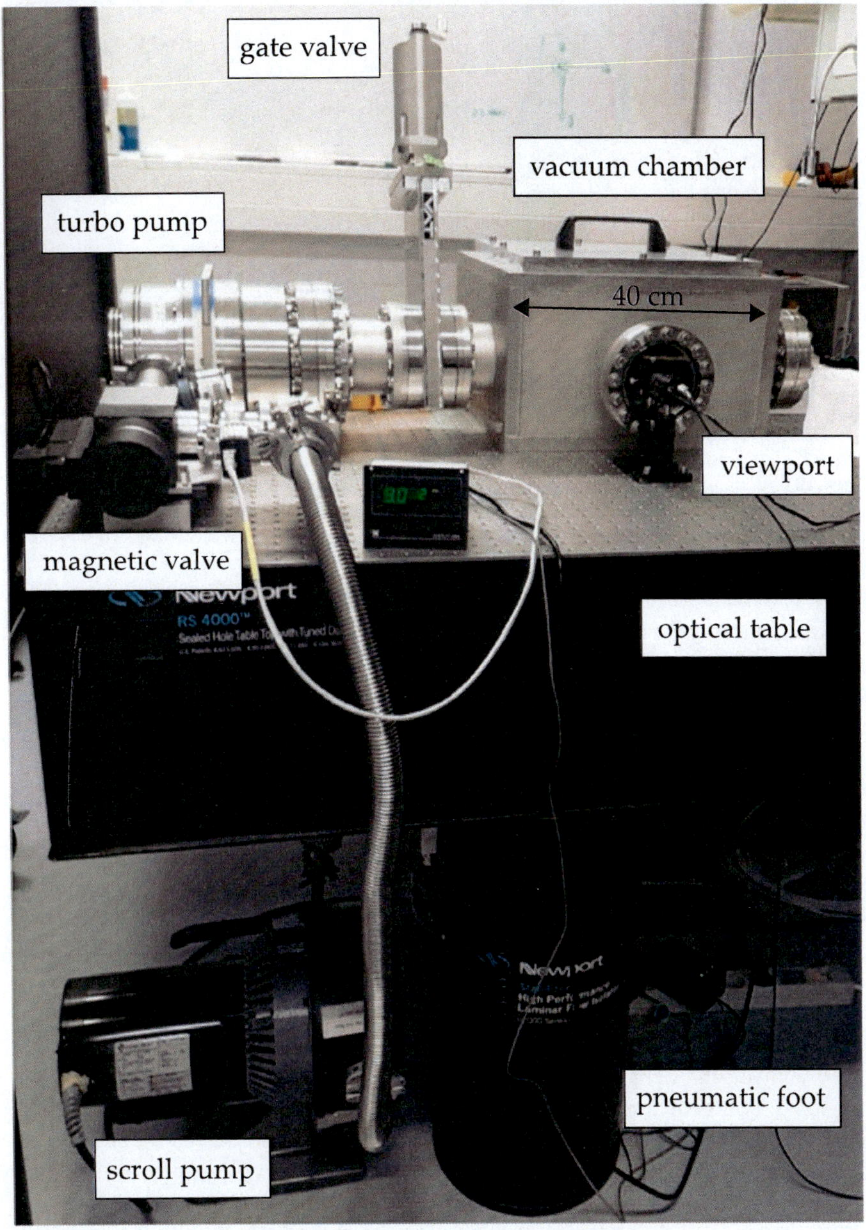

Fig. 4.9 Picture of the vacuum system comprised by the vacuum chamber, gate valve, turbo pump, magnetic valve and scroll pump. The picture also shows the optical table suspended by one of the pneumatic feet

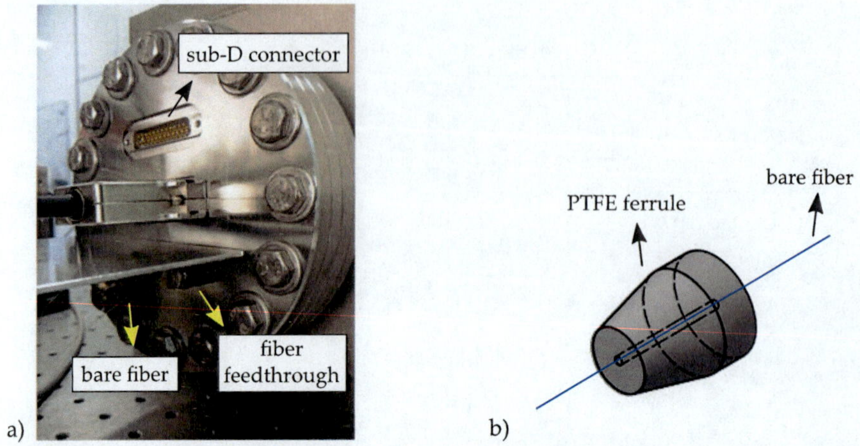

Fig. 4.10 Vacuum feed-through: **a** picture of the vacuum flange containing two sub-D connectors for the PDs, translation stage and PZT electronics, and two fiber feed-through, **b** illustration of the PTFE ferrule surrounding the bare fiber

isolation feet from the company Newport model VIB100, shown in Fig. 4.7c. The feet provided isolation above 11 Hz for vertical and above 8.5 Hz for horizontal motion. The values of these frequencies were taken from the manufacturer data sheet and they can change slightly as a function of the total breadboard load and its distribution.

Each isolation stage ideally works as a damped harmonic oscillator, meaning that below resonance the spring is effectively rigid so ground motion is completely transmitted to table/breadboard motion, while for higher frequencies the motion transfer is proportional to $1/f^2$. At resonance, the motion is amplified by a magnitude that depends on the quality factor of the system, similar to the description in Sect. 2.1.3. A calculation of the transmissibility, which is the ratio between the transmitted and the input motion [114], of the VIB 100 isolation foot is shown in Fig. 4.11 for the parameters given by the manufacturer. The transmissibility from ground to breadboard motion will be the product of the transmissibility of each isolation stage (Fig. 4.12).

To analyze the behavior of the isolation stages, the vertical displacement was measured by a geophone under three configurations: on the vacuum chamber floor when the optical table suspension was turned off (red) and on (blue), and on top of the breadboard with the suspension turned on (green). As shown by the red and blue curves, the pneumatic feet provided isolation above 10 Hz, with a maximum displacement reduction of two orders of magnitude at 25 Hz. In all measurements the vacuum chamber lid was placed at its top, but it could not be fully closed because of the geophone cables. Comparing the green curve with the blue curve, it is possible to see that the displacement noise on top of the breadboard is further reduced above 20 Hz by the VIB 100 feet. For frequencies higher than 100 Hz the residual noise is lower than the intrinsic noise of the geophone and therefore the measured

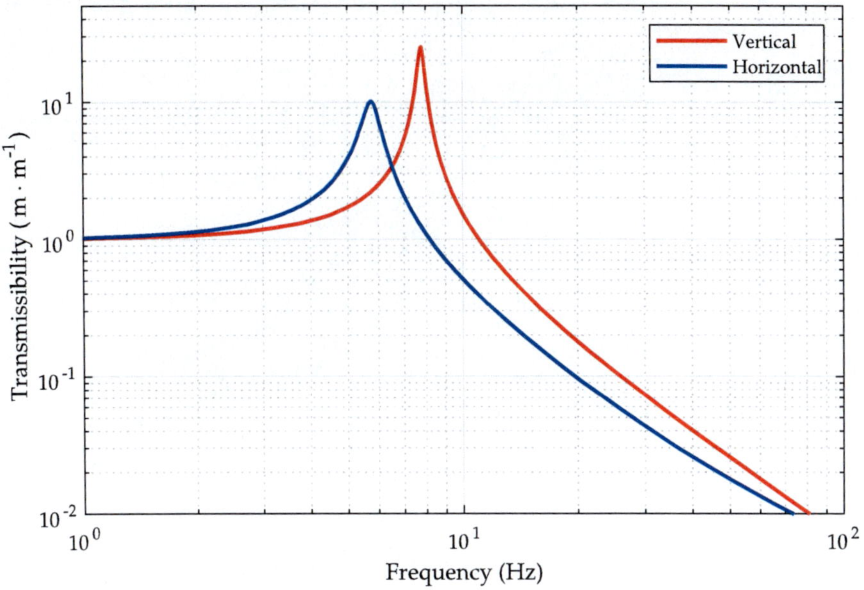

Fig. 4.11 Transmissibility of the VIB 100 isolation foot. The calculation was done assuming a vertical and horizontal resonant frequency of 7.8 Hz and 5.8 Hz respectively, and vertical and horizontal damping ratios of 0.02 and 0.05 respectively (taken from the VIB 100 data sheet)

displacement at the breadboard represents only an upper limit for vertical motion. The black dashed line shows a typical displacement noise model of a quiet site, which is $10^{-9}\text{m} \cdot \text{Hz}^{-1/2}$ and frequency independent at low frequencies, and it falls like $1/f^2$ at high frequencies [11]. The excess noise measured in the laboratory is expected since it was located in a densely populated area.

4.2.5 Michelson Interferometer Control Loop

This section describes the control loop used to lock the Michelson interferometer for displacement readout purposes. As explained in Sect. 2.2.1, the interferometer response is a nonlinear function of differential arm displacement or phase. For this reason, the calibration factor relating the interferometer output power to differential displacement depends on the operational point of the interferometer, i.e., on the constant difference between its arm lengths. Since this constant length has to be controlled on a nm scale, a feedback control loop is necessary to assure that the interferometer stays locked at the same operational point during the measurements.

The dark-fringe is the optimum operational point in order to avoid coupling of power noise from the sensing beam to the oscillator's position readout. A dark-fringe lock similar to what was described in Sect. 2.2.1 was performed on the setup. The

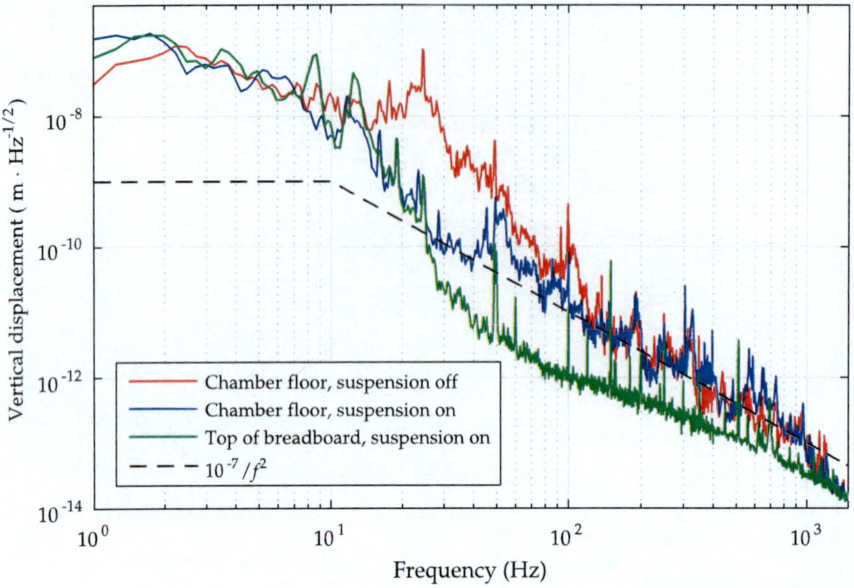

Fig. 4.12 Effect of the vibration isolation stages: ASD of vertical displacement measured by a geophone at the vacuum chamber floor with pneumatic suspension turned off (red) and on (blue), and on the top of the breadboard with suspension turned on (green). The black dashed line shows a model for a typical displacement noise of a quiet site

actuator for the differential arm length was a 17 mm long PZT with a travel range of 15 μm. The PZT was glued to the 0.5 inch end mirror of the interferometer south arm (see Fig. 4.2). With this configuration, modulations in the mirror position with frequencies above 150 kHz did not result in a sufficiently large error signal for the dark lock. This posed a problem since the modulation frequency was very close to the bandwidth in which the measurements were performed (100 kHz), and power noise at this frequency can still couple in the interferometer sensitivity. For this reason, the mid-fringe lock presented a higher signal to noise ratio and was the operational point used for all the measurements shown in this thesis. To reduce power noise coupling from the sensing beam via the mid-fringe, a traditional power stabilization control loop was used, shown in Fig. 4.2. The in-loop photodetector PS PD was placed outside the vacuum chamber, before the fiber coupling, and the power actuator used was a fast amplitude modulator port directly at the NPRO laser. The open loop unity gain frequency was kept at 1.5 MHz.

A diagram of the mid-fringe control loop is shown in Fig. 4.13, where each major component is depicted by a block with a specific complex transfer function represented by the italic capital letter. The control loop works as follows: the sum of the displacement caused by the transfer beam radiation pressure (x_{rp}) and by disturbances (x_{d}) is sensed by the interferometer (H_1) and the MI PD (H_2), that have a combined transfer function of $H = H_1 \cdot H_2$. The sensor signal V_{s} is subtracted from

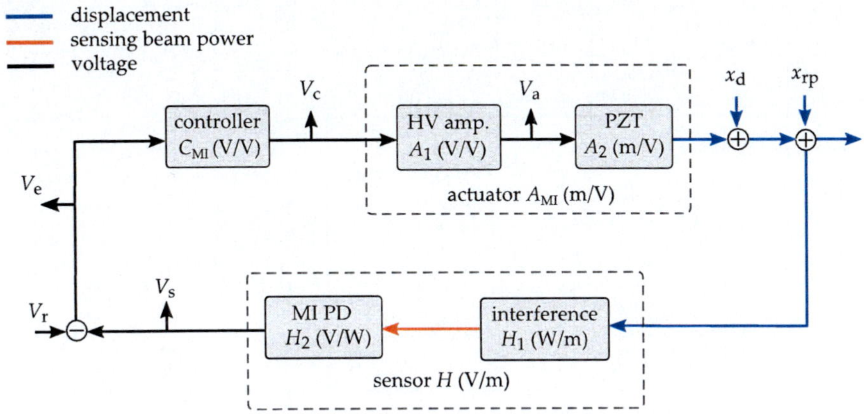

Fig. 4.13 Block diagram of the Michelson interferometer mid-fringe control loop. The radiation pressure displacement x_{rp} and the external disturbances x_d are sensed by the interferometer (H_1) and the MI PD (H_2). The sensor signal V_s is compared to the mid-fringe reference voltage V_r, and the error signal V_e is sent to the controller (C_{MI}). The control signal V_c is sent to the high voltage amplifier (A_1) and the actuator signal (V_a) is sent to the PZT (A_2) that keeps the interferometer locked at the mid-fringe

the mid-fringe reference voltage V_r and the resulting error signal V_e is amplified by a controller with transfer function C_{MI}. The control signal V_c is sent to a high voltage (HV) amplifier (A_1), and finally the actuator signal V_a is sent to the PZT (A_2) that, combined with the high voltage amplifier, has a transfer function $A_{MI} = A_1 \cdot A_2$. The PZT converts the received voltage into mirror displacement, keeping the interferometer locked at the mid-fringe.

The transfer functions of the individual blocks and the open-loop transfer function $G_{MI} = H \cdot C_{MI} \cdot A_{MI}$ are shown in Fig. 4.14. The controller (green curve) consisted of an adjustable proportional stage and an integrator stage with a gain of 40 dB at 1 Hz for a unity gain frequency (UGF) of 130 Hz. The HV amplifier (blue curve) was essentially a low pass filter with a gain of 25 dB and a corner frequency of 850 Hz, chosen to avoid electronic noise coupling into displacement at frequencies above the open-loop UGF. The PZT transfer function, not shown in the plot, is expected to be flat up to 700 Hz, and it had a magnitude of $A_2 = 12.43\ \mu\text{m} \cdot \text{kV}^{-1}$. The magnitude was determined by measuring the necessary voltage to scan the PZT mirror over a full interferometer fringe, which corresponds to a mirror displacement of $\lambda/2$. The MI PD transfer function (not shown in the figure), is flat up to the MHz regime and it had a magnitude of $H_2 = 1.28\ \text{V} \cdot \text{mW}^{-1}$. According to Eq. 2.34, the sensor transfer function is expected to be frequency independent and with a magnitude given by:

$$H = \frac{2\pi}{\lambda}(V_{max} - V_{min}), \tag{4.1}$$

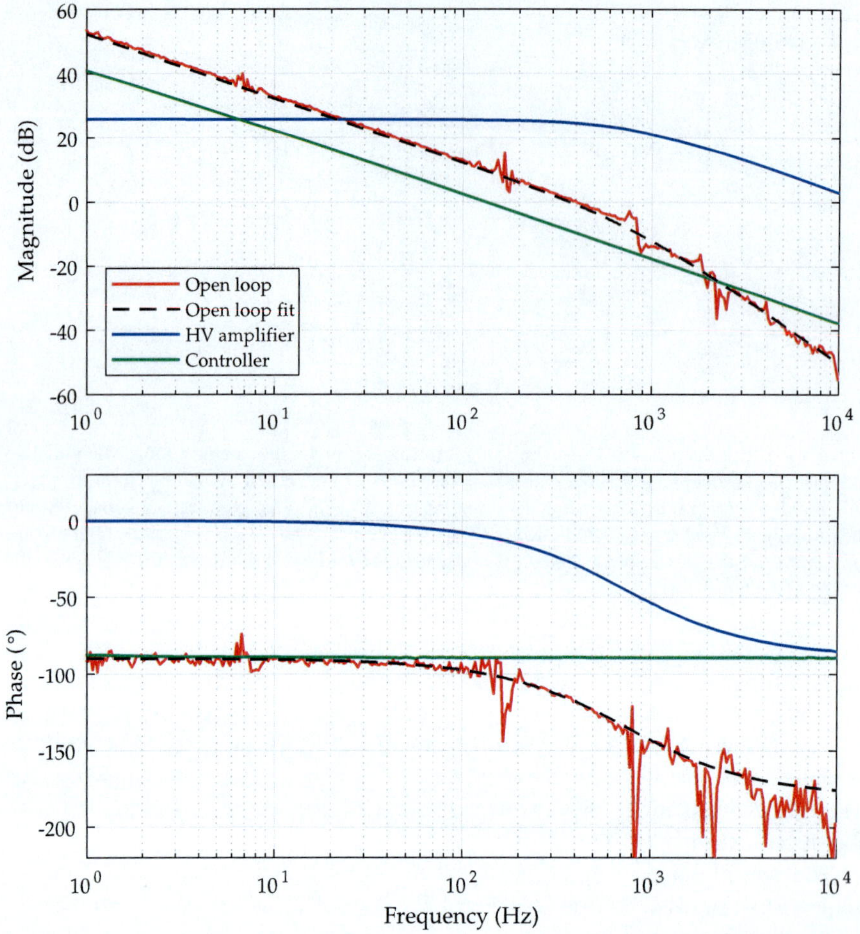

Fig. 4.14 Bode plot of the measured transfer functions of the Michelson interferometer mid-fringe lock control loop. The dashed black curve is a fit to the measured open loop transfer function

where V_{max} and V_{min} are the maximum and minimum output voltage measured by the MI PD on a full interferometer cycle. This transfer function depends on the sensing beam mean power and also on the interferometer contrast, and therefore it was measured for each individual configuration. The reference mid-fringe voltage V_r corresponds to the value $(V_{max} - V_{min})/2$ and it was adjusted in the controller.

The open-loop transfer function (red curve) was measured from the error to the sensor signal by injecting a swept sine in the error signal with the loop closed (to assure the interferometer was kept in the mid-fringe during the measurement). The open-loop UGF was set to 410 Hz in order to provide sufficient gain at the micro-oscillators fundamental longitudinal resonance frequency (below 230 Hz) and keep the interferometer locked at the mid-fringe.

The details of how the micro-oscillator displacement was obtained from the control loop output signals will be discussed in Sect. 5.1.1, which also presents a full analysis of electronic noise coupling into the displacement readout.

4.3 Interferometer and Micro-oscillator Alignment Procedure

This section describes the procedure to align the interferometer and the micro-oscillators. To avoid exposure of the oscillators to dust and also for simplicity, a half inch mirror was used in the translation stage during the interferometer alignment. An adapter was used to guarantee that the mirror reflective surface is roughly at the same plane as the reflective surfaces of the oscillators mirror pads. The procedure for the interferometer alignment was:

1. First the 50 mm lens was positioned in the center of the incoming sensing beam, in order to avoid distortions and shifts in the beam when moving the lens longitudinally, as needed in a later step. Then, the cube beamsplitter was positioned such as the reflected beam is 90° with respect to the incoming beam.
2. In the second step the PZT and translation stage mirror mounts were placed 2 cm away from the reflective surface of the beamsplitter, and the pitch and yaw of the mounts were adjusted so that the reflected beams overlap with the incoming beam. This adjustment is important to ensure that during scanning, the PZT mirror just translates the beam longitudinally, without introducing horizontal or vertical shifts.
3. The fine alignment of the yaw and pitch of the translation stage was made by scanning the PZT mirror and maximizing the interferometer contrast. Since this is a confocal interferometer, it is crucial that the macroscopic length of the two arms are equal otherwise the beams will interfere with different beam sizes and a perfect overlap will not be possible. To match the arm length, the translation stage was moved longitudinally until there were no more concentric rings in the CCD camera (only transition between dark and bright spots) and the contrast was maximized. The final contrast was then 99.4%.
4. Finally, the longitudinal position of the 50 mm lens was adjusted such as the beam waist is located at the plane of the cantilevers. The waist was measured using a beam profiler consisting of a razor blade mounted on an adapter placed in the translation stage. The measured horizontal and vertical waist were 15.85 μm and 15.90 μm, assuming a step size of 30 nm of the translation stage.

After aligning the interferometer, the sensing and transfer beams were aligned to each other such that they overlap at the micro-oscillator plane (guaranteeing that they will hit the same spot on the micro-oscillator). The procedure for this alignment, shown in Fig. 4.15, was:

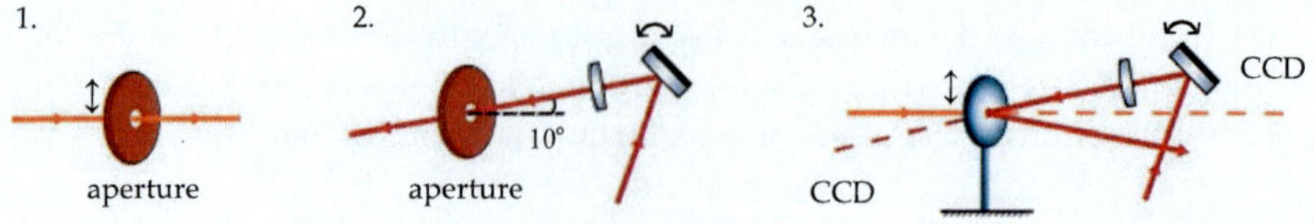

Fig. 4.15 Illustration of the procedure for the pre-alignment of the transfer and sensing beams at the micro-oscillator plane

1. First, an aperture with a radius of 100 μm was placed in the translation stage and was adjusted to be centered to the sensing beam.
2. Then, the transfer beam was aligned to the center of the aperture with an horizontal angle of incidence around 10°. This was the minimum angle that gave clearance for the transmitted sensing beam (which is later used for alignment) and this configuration was also chosen as an easier way to separate the incoming and reflected beams.
3. After this pre-alignment, the chip containing the micro-oscillators was placed and the sensing and transfer beams were pre-aligned to a specific oscillator. This alignment was done by observing the shadows in the transmitted light from the micro-oscillators in the CCD cameras. The micro-oscillators were identified by translating the chip horizontally and vertically, and counting the shadows following the layout shown in Fig. 4.2b.

To understand how the transfer and sensing beam were fine adjusted to the center of the micro-oscillator's mirror pad it is important to consider a realistic model that includes higher mechanical modes expected in the oscillators. Figure 4.16 shows the first three mechanical modes expected for the oscillators: (a) longitudinal mode—the mirror pad and the cantilever are displaced longitudinally and in phase with each other, (b) yaw mode—mirror pad and cantilever rotate horizontally with respect to the undisturbed position, and (c) pitch mode—mirror pad rotates vertically with respect to the cantilever. From the oscillator's symmetry, the effective mass of the yaw mode is expected to be higher than the pitch mode, and therefore the resonance frequency of the yaw mode is lower than the pitch mode. The complete oscillator motion can be mathematically represented by a continuous series of coupled harmonic oscillators which are coupled with each other by joint spring constants. Then, the longitudinal motion x of a certain point on the mirror surface, e.g. the one hit by the laser beam, can be described as the motion of a single system formed by independent damped harmonic oscillators each having its own effective mass m_{m}, resonance frequency Ω_{m}, and loss angle ϕ_{m} [115]:

$$\frac{x(\Omega)}{F_{\mathrm{ext}}(\Omega)} = \sum_{\mathrm{m}} \frac{1}{-m_{\mathrm{m}}\Omega^2 + m_{\mathrm{m}}\Omega_{\mathrm{m}}^2(1 + i\phi_{\mathrm{m}})} . \tag{4.2}$$

If the transfer beam is centered to the mirror pad, then the radiation pressure driving should ideally not excite the yaw and pitch modes of the oscillator. By

Fig. 4.16 Schematic of the four first mechanical modes expected for the micro-oscillators (illustration taken from reference [73])

using this reasoning, the transfer beam was centered to the mirror pad by reducing the yaw and pitch resonances in a radiation pressure transfer function measurement. This measurement was performed by applying a swept sine function to the AOM and measuring the oscillator displacement with the interferometer. For this measurement, the sensing beam was intentionally off-centered from the mirror pad, to maximize the interferometer sensitivity to the pitch and yaw motion. Finally, the sensing beam was aligned to the center of the mirror pad by minimizing the yaw and pitch peaks in the thermal noise displacement, since the interferometer should not be sensitive to the high order resonances in this situation (assuming only negligible effects in the interferometer readout caused by the small rotations of the reflected beam). During this step, the transfer beam was kept at the same power as the previous step, in order to keep the oscillator at the same mean, static position. Since the steering mirrors for the sensing and transfer beams were not motorized, this alignment was made in air and it was assumed that misalignments between the sensing and transfer beam when evacuating the chamber were small enough for the purposes of this experiment. In fact, the contrast of the interferometer remained the same at different pressures on the chamber, which is an indicator for small or no misalignment of the beams under vacuum. However, in a future generation of the setup it would be better to implement one motorized mirror mount to guarantee a precise alignment also in high vacuum. This is specially important when increasing the power in the transfer beam, as it will be discussed later.

After these steps, the beams are aligned to each other at the mirror pad plane. Other oscillators were aligned to the sensing and transfer beams by translating the chip mount and reducing the yaw and pitch peaks in the thermal noise displacement measurement, as shown in Fig. 4.17. The pitch and yaw resonances in the interferometer displacement noise (red curve) completely vanished when the oscillator was centered to the transfer and sensing beams (blue curve). Both curves were taken at a pressure of 10^{-2} mbar. The figure also shows harmonic peaks of the individual modes, and also peaks representing a beat between different modes. Small drifts in the oscillator alignment were observed when the pressure on the chamber was changed, or after a long period of time. For this reason, the oscillator was realigned[3]

[3] The chip had to be translated, on average, around 6 μm in vertical and horizontal directions in the realignment procedure.

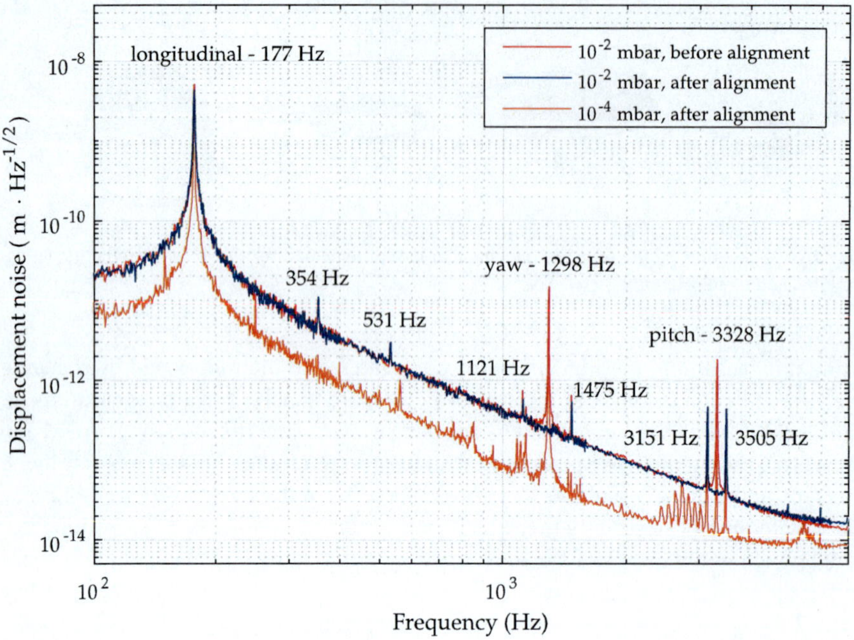

Fig. 4.17 Micro-oscillator alignment: ASD of the interferometer displacement noise measured with the micro-oscillator LA7 at 10^{-2} mbar before (red) and after (blue) optimizing the oscillator alignment, and at 10^{-4} mbar after optimizing alignment (orange). The following peaks are marked in the blue curve: longitudinal fundamental resonance (177 Hz), second (354 Hz) and third (531 Hz) longitudinal harmonics; yaw resonance (1298 Hz), beat between yaw and longitudinal (1121 and 1475 Hz); pitch resonance (3328 Hz) and beat between pitch and longitudinal (3151 and 3505 Hz)

such as the peak and yaw fundamental resonances would not couple at the thermal noise displacement measurement. At pressures lower than 10^{-4} mbar, the realignment turned out to be a slow process due to the long settle time of the micro-oscillator, which was excited by the translation stage step motion. In addition to that, the alignment was also very sensitive, and one step of the translation stage was not always sufficiently precise to make the peaks vanish. For such pressures, small alignment drifts were observed, which brought the pitch and yaw peaks back to higher values. Due to these reasons, some of the measurements in this thesis will still shown pitch and yaw resonances.

Chapter 5
Power Stabilization Via Radiation Pressure—Experimental Results

Abstract This chapter presents the results from the power stabilization experiment, whose setup was described in Chap. 4. The experiment was divided into three parts: sensing noise, power sensing, and power stabilization. In the sensing noise part, the sources of displacement noise in the Michelson interferometer readout were analyzed. In order to avoid the high thermal noise from the micro-oscillators, the technical noise sources were investigated and reduced in a configuration were the micro-oscillator chip was substituted by a rigidly fixed 0.5 in mirror. Afterwards, the interferometer displacement noise was measured and analyzed for different micro-oscillators. In the second part, the interferometer containing the micro-oscillator was used as a sensor for the free running power noise of the transfer beam. A measurement of the optomechanical response of the micro-oscillator was also performed. Finally, the transfer beam power stabilization control loop was enabled, and the power stability performance of the out-of-loop beam was measured for different transfer beam powers.

5.1 Sensing Noise: Interferometer Sensitivity with a Fixed Mirror

As discussed in the previous chapters, the Michelson interferometer containing the micro-oscillator forms the in-loop sensor for the power stabilization scheme via radiation pressure. Hence, noise sources coupling to the interferometer readout will be imprinted on the out-of-loop beam by the stabilization control loop, and will set a limit to the maximum power stability that can be achieved with this scheme. For this reason, it is crucial to minimize the interferometer technical noise down to the stabilization requirements of the experiment. In this thesis, the requirements were set by the expected structural thermal noise from the micro-oscillators at room temperature. This thermal noise contribution could be reduced by operating the experiment with cryogenic temperatures. However, due to the increase in complexity and also

M. Trad Nery, *Laser Power Stabilization via Radiation Pressure*, Springer Theses,
https://doi.org/10.1007/978-3-030-95868-8_5

the non-availability of a cryostat, this proof-of-principle experiment was operated at room temperature.

To allow a clear identification of technical noise sources, the micro-oscillator chip was substituted by a 0.5 inch mirror which was rigidly fixed in the translation stage. This configuration was necessary to avoid the high thermal noise contribution of the micro-oscillators, especially from viscous damping at atmospheric pressure (necessary at this point to facilitate changes in the setup), which would mask the contribution from technical noise sources. Another reason for this configuration was to identify noise sources that need be lowered in a future generation of this experiment, when thermal noise is reduced.

In the next sections, possible technical noise sources and their coupling paths are discussed. Finally, the best interferometer sensitivity is compared with the expected thermal noise of different micro-oscillators.

5.1.1 Electronic Noise

Electronic noise from the MI control loop can couple into the displacement readout via different paths, as shown by the block diagram in Fig. 5.1. The displayed noise sources are: sensor noise N_s (dark noise of the MI PD), reference noise N_r, controller noise N_c, and actuator noise N_a (HV amplifier noise). They originate from the intrinsic noise of the individual electronic components such as operational amplifiers, resistors, and capacitors. These noise sources are summed, amplified, or even reduced according to the transfer function of each stage inside the electronics block. For this reason, they will be treated as a single noise source that couples at the input

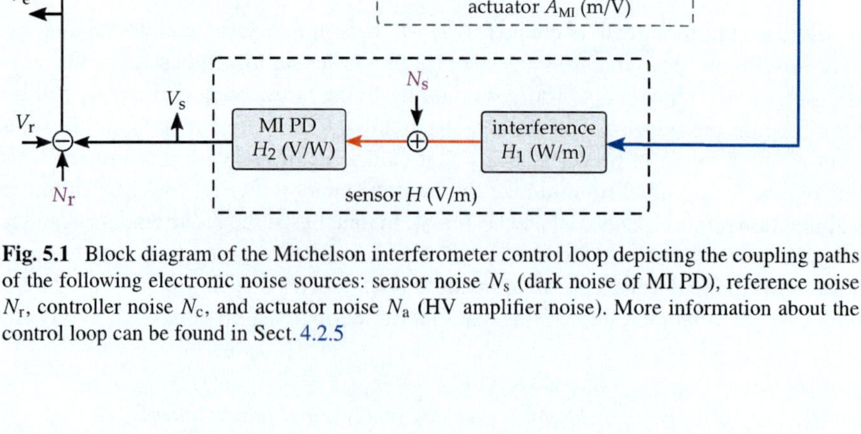

Fig. 5.1 Block diagram of the Michelson interferometer control loop depicting the coupling paths of the following electronic noise sources: sensor noise N_s (dark noise of MI PD), reference noise N_r, controller noise N_c, and actuator noise N_a (HV amplifier noise). More information about the control loop can be found in Sect. 4.2.5

of each block, although the reader should keep in mind that, experimentally, the noise is usually measured at the output of the electronics.

The readout of the total interferometer displacement $x_{rp} + x_d$ can be performed by projecting different output signals from the control loop to displacement. To decide which signal is more appropriate for the projection, the coupling of electronic noise must be taken into account. When the loop is closed and working in a steady state regime, the controlled output displacement x_0 can be calculated from the block diagram as [116]:

$$x_0 = \frac{x_d + x_{rp} + [(-N_s \cdot H_2 + N_r + N_c) \cdot C_{MI} + N_a] \cdot A_{MI}}{1 + G_{MI}}, \tag{5.1}$$

where $G_{MI} = H \cdot C_{MI} \cdot A_{MI}$ is the open-loop transfer function.

The relation between the sensor signal V_s and the total displacement can be obtained by substituting Eq. 5.1 into the expression $V_s = x_0 \cdot H + N_s \cdot H_2$, which leads to:

$$V_s = \frac{(x_{rp} + x_d) \cdot H}{1 + G_{MI}} + \frac{N_s \cdot H_2}{1 + G_{MI}} + \frac{G_{MI}}{1 + G_{MI}} \left[N_r + N_c + \frac{N_a}{C_{MI}} \right]. \tag{5.2}$$

Hence, the projection of the sensor signal to displacement x_{V_s} (in meters) is:

$$x_{V_s} = \frac{V_s \cdot (1 + G_{MI})}{H} \tag{5.3}$$

$$= x_{rp} + x_d + \frac{N_s}{H_1} + C_{MI} \cdot A_{MI} (N_r + N_c) + N_a \cdot A_{MI}. \tag{5.4}$$

Similarly, the relation for the actuator signal is obtained by substituting Eq. 5.1 into the expression $x_0 = x_{rp} + x_d + V_a \cdot A_2$. This leads to:

$$V_a = -\frac{(x_{rp} + x_d) \cdot G_{MI}}{A_2 \cdot (1 + G_{MI})} + \frac{A_1}{1 + G_{MI}} [(-N_s \cdot H_2 + N_r + N_c) \cdot C_{MI} + N_a]. \tag{5.5}$$

The projection of the actuator signal to displacement x_{V_a} (in meters) is then:

$$\begin{aligned} x_{V_a} &= \frac{V_a \cdot A_2 \cdot (1 + G_{MI})}{G_{MI}} \\ &= -x_{rp} - x_d - \frac{N_s}{H_1} + \frac{N_r + N_c}{H} + \frac{N_a}{H \cdot C_{MI}}. \end{aligned} \tag{5.6}$$

As expected, Eqs. 5.4 and 5.6 show that electronic noise couples in different ways when performing the displacement readout via the sensor and actuator signals. For frequencies below the UGF, where $G_{MI} \gg 1$, electronic noise dominates the sensor signal, while it is suppressed in the actuator signal. For high frequencies, where $G_{MI} \ll 1$, the opposite situation occurs. Therefore, it is common that the interferometer displacement is obtained as a combination between the projected actuator signal for frequencies below UGF, and the projected sensor signal for frequencies above UGF.

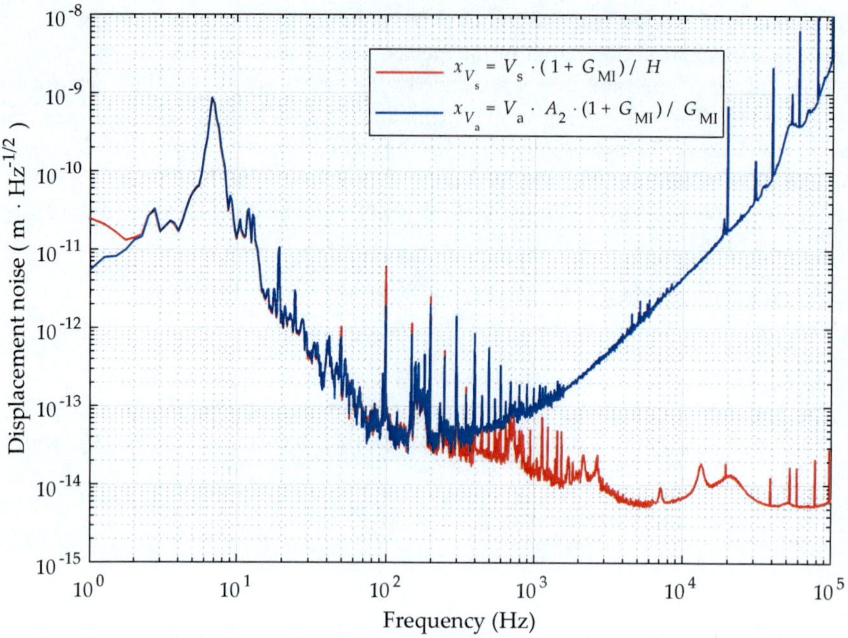

Fig. 5.2 Comparison between the interferometer displacement noise obtained from the projection of the sensor signal x_{V_s} (red) and of the actuator signal x_{V_a} (blue). The MI open loop unity gain frequency was set to 200 Hz

Previous investigations on the setup showed that the interferometer sensitivity was electronic noise limited for frequencies above 1 kHz. To mitigate this noise, the following changes were made in the electronics: implementation of operational amplifiers with lower noise, reduction of the proportional gain in the controller, increase in the sensing beam power, avoidance of high resistance resistors, and implementation of a low pass filter in the HV amplifier. Figure 5.2 shows a comparison between the projected displacement noise x_{V_s} and x_{V_a}, after the mentioned modifications. The curves meet for frequencies below the UGF of 200 Hz, whereas for higher frequencies the displacement obtained from the actuator signal still shows excessive noise. This behavior means that electronic noise below UGF does not limit the interferometer sensitivity. Therefore, all displacement curves shown in this thesis were plotted only with the projection of the sensor signal (absolute value of Eq. 5.3). Additionally, the sensor signal was used as the error signal for the power stabilization via radiation pressure, which is discussed in Sect. 5.4.1.

To understand how the remaining electronic noise could still limit the interferometer sensitivity, a measurement of the displacement noise was compared with projections of the sensor, reference, controller, and actuator noise, according to Eq. 5.4. The results are shown in Fig. 5.3. The HV amplifier electronic noise was measured directly at the output of the electronics, instead of the usual monitor port which divided the output signal by a factor of 100. Except for frequencies below 2 Hz, where the projections of the controller and reference noise are high, electronic noise

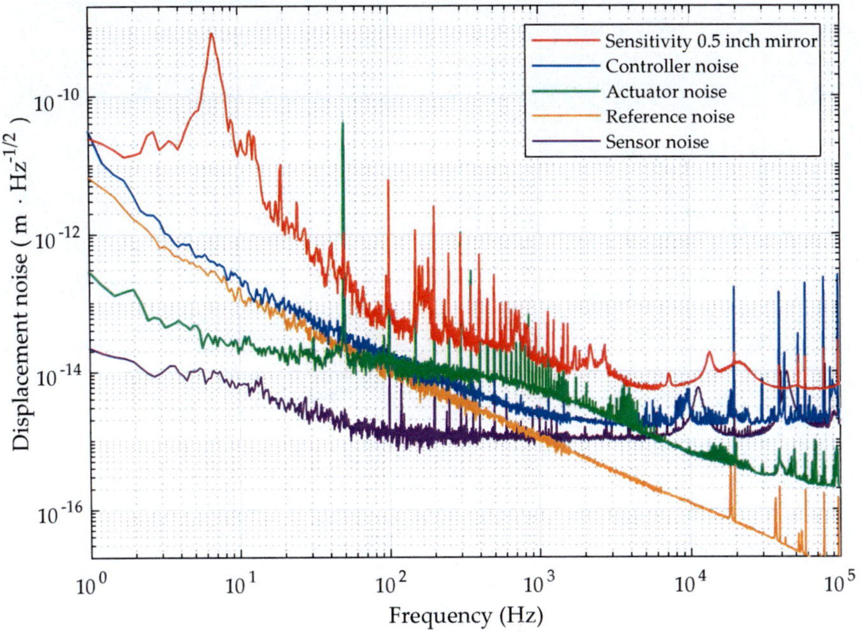

Fig. 5.3 Interferometer displacement noise measured with the 0.5 inch mirror (red), and projections of electronic noise to displacement: controller noise (blue), actuator noise (green), reference noise (yellow), and sensor noise (purple). The projections were performed using Eq. 5.4

does not pose an impediment to optimize the interferometer sensitivity by a factor of 2. If further improvement is necessary, then the following noise reductions should be made: controller and reference noise for frequencies below 200 Hz (the UGF), actuator noise from 200 Hz to 4 kHz, and controller and sensor noise above 4 kHz. These reductions can be made by either exchanging the electronic components to lower noise counterparts, or alternatively, by changing the coupling strength of electronic noise. A straightforward way to reduce sensor noise coupling is to increase the power in the interferometer, which increases the magnitude of the transfer function H_1. If the open-loop UGF is adjusted to be the same, then the magnitude of $C_{\mathrm{MI}} \cdot A_{\mathrm{MI}}$ should be proportionally smaller, which will lower the contributions from reference, controller, and actuator noise. In this case, reducing A_{MI} over C_{MI} might be preferred since the contribution from actuator noise depends only on the magnitude of A_{MI}.

5.1.2 *Laser Frequency Noise*

Laser frequency noise couples into displacement noise via an inequality between the macroscopic arm lengths of the interferometer. In this situation, the laser frequency noise $\delta\nu$ at the interferometer input experiences different delays in the interferometer arms and it does not cancel at the readout port. The induced displacement noise x_{f} is proportional to the arm length difference ΔL, and to the laser frequency noise [11]:

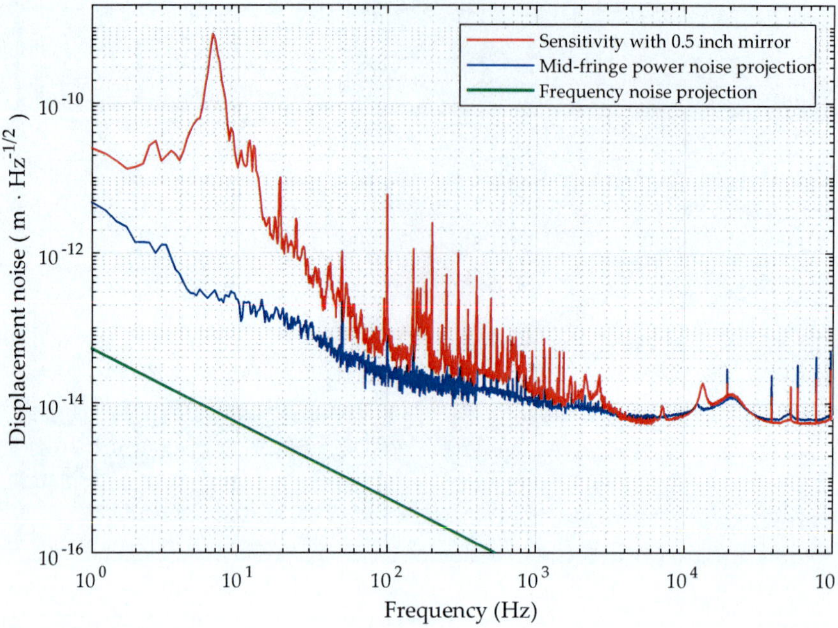

Fig. 5.4 ASD of the interferometer displacement noise (red), and sensing beam power noise coupling via mid-fringe readout (blue) and laser frequency noise (green) projection to interferometer displacement noise. The sensing beam power noise was measured by the MIPD with one of the interferometer's arm blocked

$$x_{\mathrm{f}} = 2\Delta L \cdot \frac{\delta\nu \cdot \lambda_0}{c}. \tag{5.7}$$

Since this experiment employs a confocal interferometer, ΔL should in principle not be much higher than the beam Rayleigh range of 0.75 mm (calculated for a waist of $16\,\mu\mathrm{m}$), otherwise a contrast of 99.4% would not be possible. The frequency noise projection estimated for this experiment is shown in Fig. 5.4. The projection was calculated with the typical frequency noise of an NPRO laser of $10^4/f\ \mathrm{Hz} \cdot \mathrm{Hz}^{-1/2}$ [5], and with $\Delta L = 0.75\,\mathrm{mm}$. The plot shows that the frequency noise contribution is more than 2 orders of magnitude below the interferometer displacement noise, and therefore it was not a limiting noise source for this experiment.

5.1.3 *Sensing Beam Power Noise*

Since the interferometer is locked to the mid-fringe, power noise from the sensing beam will couple directly at the interferometer readout. This coupling would not be an issue if the free running power noise of the sensing and transfer beam at the fiber outputs were coherent with each other. In that case, a power actuator could be used before the splitting of the beams to simultaneously stabilize the power of the transfer and sensing beam using the stabilization via radiation pressure control loop. In this experiment, however, the coherence between the power noise in the beams was lost at the output of the fibers. This could have been caused, for example, by beam jitter at the fiber input or by polarization changes in the fiber.

The projection of power noise to displacement noise $x_{\mathrm{p,mf}}$ via the mid-fringe readout can be obtained from the interferometer transfer function H_1:

$$
\begin{aligned}
H_1 &= \frac{2\pi(P_{\mathrm{max}} - P_{\mathrm{min}})}{\lambda}, \\
x_{\mathrm{p,mf}} &= \frac{\mathrm{RPN_s} \cdot (P_{\mathrm{max}} - P_{\mathrm{min}})}{2H_1} = \frac{\mathrm{RPN_s} \cdot \lambda}{4\pi},
\end{aligned}
\tag{5.8}
$$

which only depends on the sensing beam relative power noise $\mathrm{RPN_s}$. In order to lower this contribution, a traditional power stabilization was realized, as discussed in Sect. 4.2.5. The open loop UGF achieved with this scheme was approximately 1 MHz and, due to saturation problems in the controller electronics, it could not be further increased. The residual sensing beam power noise was not significantly improved in comparison to the configuration when the NPRO noise eater was turned on. Figure 5.4 compares the interferometer displacement noise and the noise projection $x_{\mathrm{p,mf}}$ when the NPRO noise eater was turned on. It is clear that power noise limits the interferometer sensitivity for frequencies above 100 Hz. Hence, improvements on the power noise coupling via the mid-fringe should be a priority in a future generation of this experiment.

Another coupling path of sensing beam power noise, which is present with the micro-oscillator mirror, is via radiation pressure noise. The induced displacement noise $x_{\mathrm{p,rp}}$ in this case depends on the mechanical susceptibility of the oscillator, and on the absolute sensing beam power noise:

$$
x_{\mathrm{p,rp}} = \frac{\overline{P}_{\mathrm{s}} \cdot \mathrm{RPN_s}}{mc\sqrt{\left(\Omega_0^2 - \Omega^2\right)^2 + \Omega_0^4\phi^2}}, \tag{5.9}
$$

where it was considered that half of the sensing beam power hits the micro-oscillator. In order to reduce this coupling, it is important to keep the sensing beam mean power on the lowest level necessary to overcome electronic noise. Figure 5.5 compares the expected structural thermal noise for a micro-oscillator with mass of 40 ng and fundamental resonance frequency of 117 Hz, with the radiation pressure noise projection considering a sensing beam power at the input of the interferometer of 20 mW. For oscillators with a lower mechanical susceptibility, the difference between the thermal noise and the radiation pressure noise contributions will be even larger. This, together with the fact that the sensing beam power at the input of the interferometer was usually lower than 20 mW, leads to the conclusion that radiation pressure noise will not limit substantially the sensitivity of this experiment.

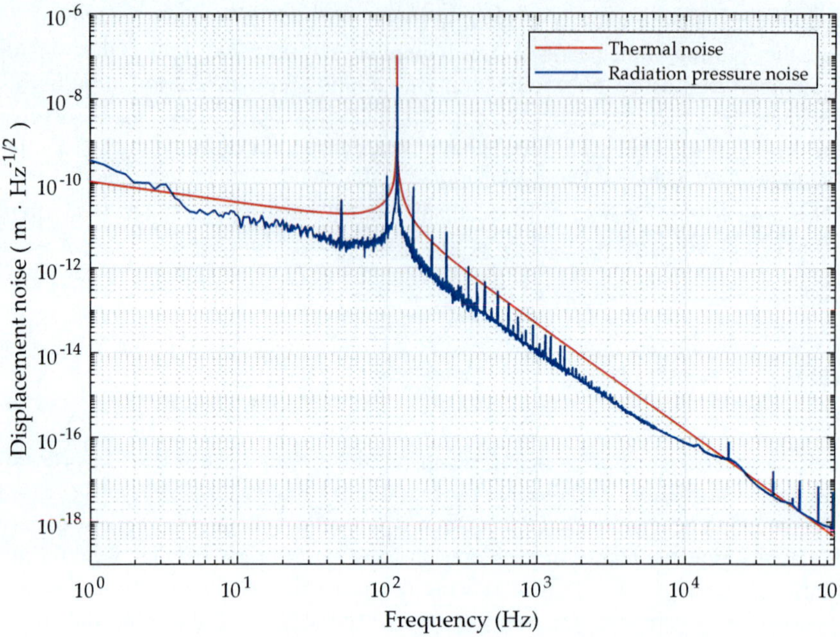

Fig. 5.5 Comparison between the expected structural thermal noise displacement at room temperature for a micro-oscillator with mass of 40 ng, longitudinal resonance frequency of 117 Hz, and quality factor of 10^4 (red curve), and the radiation pressure noise projection from the measured sensing beam power noise, considering a mean power at the input of the interferometer of 20 mW (blue curve)

5.1.4 Vibrational Noise

Vibrations are a main contributor to interferometer displacement noise x_{d}, especially at frequencies below 100 Hz. They encompass a set of noises with different origins such as seismic motion caused by ocean waves or human activities, and acoustics induced by air conditioning, ventilators and structural vibrations.

As described in Sect. 4.2.4, the interferometer was mounted on top of two passive isolation stages which considerably reduced the vibrational motion on top of the breadboard. To investigate how residual vibrations affect the interferometer sensitivity, a coherence measurement between optical table vertical motion, performed with a geophone, and interferometer displacement noise was taken. The result is shown in Fig. 5.6. For this measurement, the interferometer was placed inside the chamber with a pressure of 10^{-2} mbar. The geophone, however, was placed outside the vacuum chamber and on top of the optical table, since there was no feed-through in the chamber for the geophone cable. A significant coherence for frequencies below 1 kHz can be observed. This means that residual vertical (and possible horizontal)

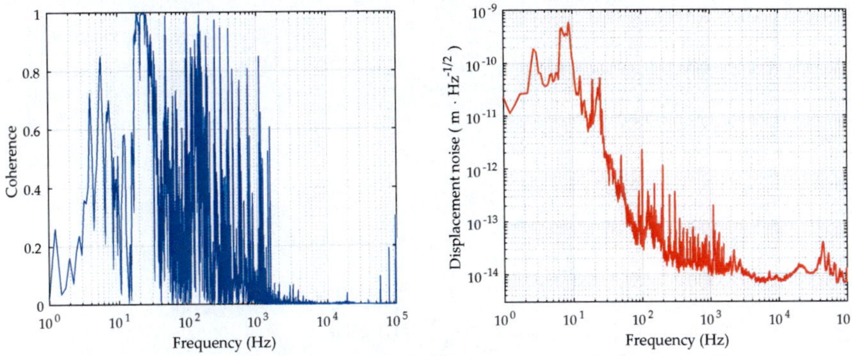

Fig. 5.6 Left: coherence measurement between vertical motion of the optical table and interferometer displacement noise. The measurement was performed with a geophone placed on top of the optical table, outside the vacuum chamber, and with the interferometer inside the chamber under a pressure of 10^{-2} mbar. Right: ASD of the interferometer displacement noise, taken simultaneously with the coherence measurement

vibrational noise still couples to the interferometer displacement noise. Hereafter, the possible paths for this coupling are discussed and investigated.

Effect of the Optical Table Pneumatic Isolation Stage

As shown in Fig. 5.7, the use of two isolation stages was important for the setup since by turning the optical table pneumatic stage on, the interferometer displacement noise is reduced between 10 and 100 Hz. This result qualitatively agrees with the measurements showed in Fig. 4.12, where a reduction of vertical motion of approximately one order of magnitude was observed at 30 Hz by turning the table suspension on. This fact points out that the interferometer sensitivity could still be limited by vibrational noise coupling via the optical table for frequencies below 100 Hz. Hence, an isolation stage with a lower resonance frequency, which can provide a higher noise suppression below 100 Hz, would be beneficial for a future configuration of the setup. The additional noise at frequencies above 1 kHz observed in this measurement was not well understood.

Direct Coupling at the Mirror Mounts

In long baseline interferometers such as LIGO, the differential arm length is directly affected by seismic noise, since the driving motion is different at the end mirror locations. Here, an opposite scenario takes place since the interferometer is compact (2 cm arm length), and sits on top of a rigid breadboard. This means that, ideally, horizontal and vertical residual motion of the breadboard should not couple in the differential mode of the interferometer for frequencies below the resonances of the optical mounts, which should be in the kHz regime or above. However, in reality the breadboard and the mounts are not fully rigid and since the PZT mirror and the translation stage mount are considerably different, there is a possibility that this direct coupling spoils the sensitivity of the interferometer. For this reason, the springs of the PZT mirror mount were elongated to their maximum tension, which

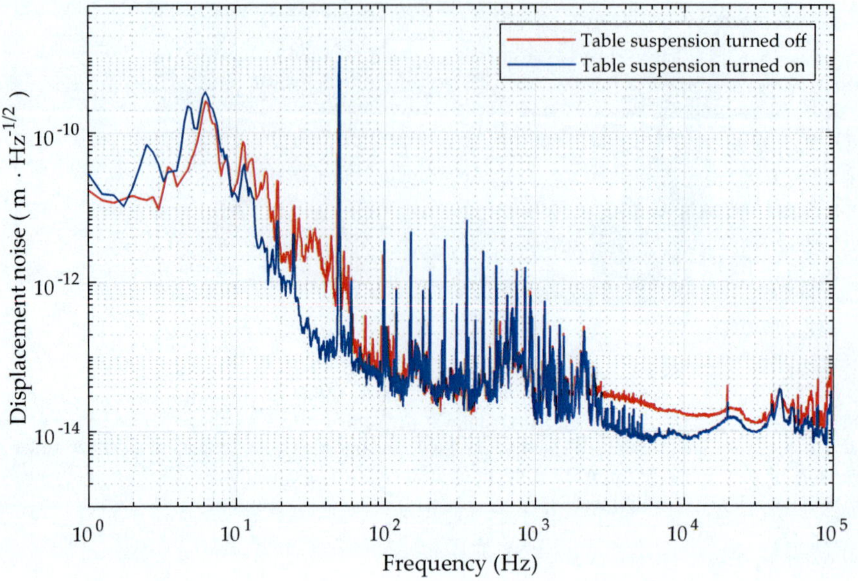

Fig. 5.7 Effect of the optical table pneumatic suspension on the ASD of interferometer displacement noise. In both measurements the breadboard was placed inside the vacuum chamber with atmospheric pressure

shifted the mount resonance to a higher frequency. No changes were observed in the interferometer sensitivity after this procedure, which fits to the expectation that the mount resonance is above the kHz regime, where the sensitivity is not limited by vibrations.

Coupling via Power Noise

Another coupling path for vibrations is via power noise at the input and output of the interferometer. Vibrations in the laser preparation area, for example, cause beam jitter at the input of the optical fibers that is converted to power noise at the input of the interferometer. Because the breadboard is connected to the vacuum chamber floor by the protective aluminum tubes surrounding the bare fibers, vibrations in the vacuum chamber can induce polarization fluctuations via changes in the tension of the fiber. Since a PBS was placed directly at the fiber output, this polarization fluctuation can induce power noise at the interferometer input. Lastly, vibrations can also induce beam jitter at the MI PD and cause an apparent power noise via the non-homogeneity in the photodiode response. To test these possibilities, the the power noise of the sensing beam inside and outside the vacuum chamber was measured with the table suspension turned on and off. No significant changes were observed in these measurements. Hence, the additional displacement noise observed in Fig. 5.7 when the table suspension was turned off seems not to be coupling via power noise.

Scattered Light

Light scattered from the nominal sensing or transfer beam paths is an indirect coupling of vibrational noise. Scattering can happen due to imperfections in the optics, dust particles, or, in the case of the micro-oscillators, bad etching and a large beam compared to the mirror pad size. Because light is scattered by one or more surfaces moving with an amplitude comparable or larger than the light's wavelength, it will acquire a differential and frequency dependent phase with respect to the nominal sensing beam, which will lead to displacement noise after interference with the sensing beam. If 1 nW of scattered light interferes with 10 mW of the nominal sensing beam at the MI PD for example, then an RPN of 10^{-4} is expected, which is enough to limit the sensitivity of the interferometer.

As stated in Sect. 4.2.1, the optical frequency of the transfer beam was shifted by 80 MHz with respect to the sensing beam frequency, in order to reduce the coupling from scattered light from the transfer beam to displacement noise. Indeed, no changes were observed in the interferometer sensitivity by turning the transfer beam on and off. However, a big candidate for scattering in the sensing beam path is the back reflected light from the interferometer mid-fringe lock. To investigate this coupling, the input and back reflected beams were intentionally misaligned from each other, but no changes were observed in the displacement noise. Scattering at the MI PD was investigated by surrounding the photodiode with a long tube to block potential side beams and also by covering the vacuum chamber viewports with black foil. No difference was observed in such a configuration. In a second second step, the interferometer was locked with a photodiode placed outside the chamber and again, the sensitivity did not change.

Acoustic Noise

Acoustic noise is highly suppressed by the vacuum chamber enclosure. This can be seen in Fig. 5.8, which compares the displacement noise when the breadboard was placed inside and outside the chamber. It was observed that the air flow boxes directly at the optical table and also in other locations of the room still had an effect in the sensitivity for frequencies between 100 Hz and 1 kHz, when the breadboard was inside the chamber. For this reason, all measurements in this thesis were done with the flow boxes turned off. This fact, together with the result of Fig. 5.7, is an indication that the observed coherence between vertical motion and displacement noise (see Fig. 5.6) from 100 Hz to 1 kHz is likely caused by residual acoustic vibrations, coupling directly at the chamber. For frequencies above 1 Hz, the setup was not limited by air inside the vacuum chamber, which could cause noise via direct acoustic coupling or via temperature changes, that alter the air refractive index and consequently the light traveled path. This was verified by an additional measurement with a pressure of 10^{-2} mbar in the chamber.

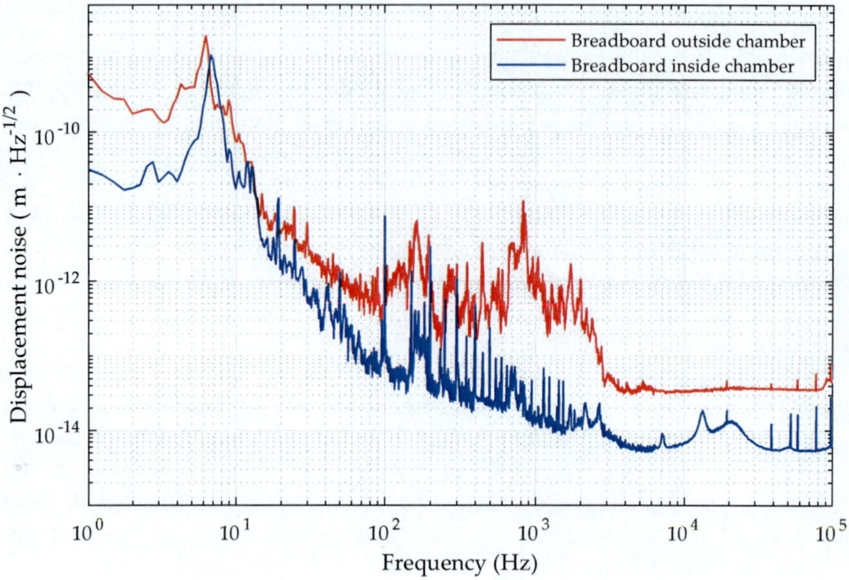

Fig. 5.8 ASD of the interferometer displacement noise measured with the breadboard outside (red) and inside (blue) the vacuum chamber with atmospheric pressure. The air flow boxes were turned off in both measurements

Beam Pointing

Beam jitter with small amplitudes at the input of the interferometer will ideally not change the interferometer contrast, but it will be transferred to jitter at the output beam of the interferometer. To investigate the importance of this effect, the beam spot size and location at the MI PD were changed, and no difference in the sensitivity was observed. During modifications on the setup, the initially plate beamsplitter was substituted by a cube one, with a different holding mount. Since no difference in the displacement noise was observed during this process, it was concluded that beam pointing introduced by the beamsplitter mount was also not an issue.

Internal Resonances of the VIB100

A common feature in the displacement noise curves shown in this thesis is a broad peak around 7 Hz. This peak was related to the vertical and horizontal resonances of the VIB isolation feet, as confirmed by the plots shown in Sect. 4.2.4. The coupling path of this noise was not fully understood, and long-term observations showed variations of the peak height that were independent of the setup configuration. This variation is illustrated in Fig. 5.9, which shows a broad change in the displacement noise curve, with a difference of one order of magnitude in the 7 Hz peak, when measured on two consecutive days without any changes in the setup in the meanwhile. Such significant change indicates that the driving of this noise is not long term stable and is likely related to changes in the vibrations of the laboratory building. In addition to that, changes in the weight distribution on the breadboard and on the

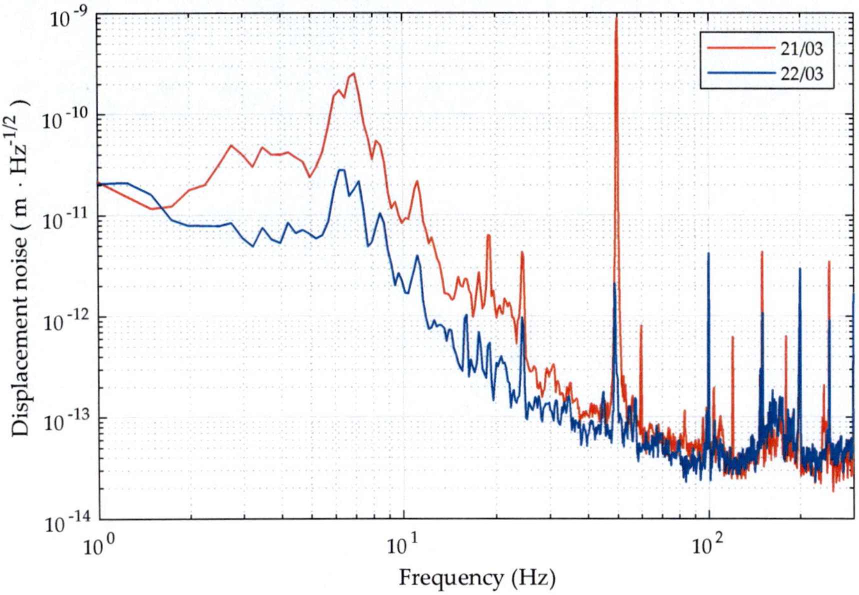

Fig. 5.9 Observation of the 7 Hz peak: ASD of the interferometer displacement noise taken on 21/03 afternoon (red), and on 22/03 morning (blue) with no changes in the setup in between. The breadboard was placed inside the vacuum chamber with atmospheric pressure

feet configuration were made in an attempt to understand the coupling and reduce the peak, but no significant changes were observed.

5.1.5 Conclusion

The best interferometer sensitivity achieved with the 0.5 inch mirror was plotted against the expected structural thermal noise of two micro-oscillators in the higher (red curve) and the lower (purple curve) limit of the mechanical susceptibility of interest in this experiment, as shown in Fig. 5.10. The figure also shows the projection of the total electronic noise and the sensing beam power noise coupling via the mid-fringe readout. The interferometer will be thermal noise limited in the frequency range between 15 Hz and a few kHz. To reach the thermal noise limit at frequencies below 15 Hz, further vibration isolation would be required, possibly with a lower corner frequency in order to avoid the high peak at 7 Hz. Due to the steep decrease in the thermal noise displacement after the longitudinal resonance frequency, the interferometer sensitivity would have to be improved by at least two orders of magnitude in order to reach a thermal noise limited performance at frequencies up to 100 kHz, which was beyond the goal of this proof-of-principle experiment. Hence, other advancements in the setup were prioritized. However, in a future generation of

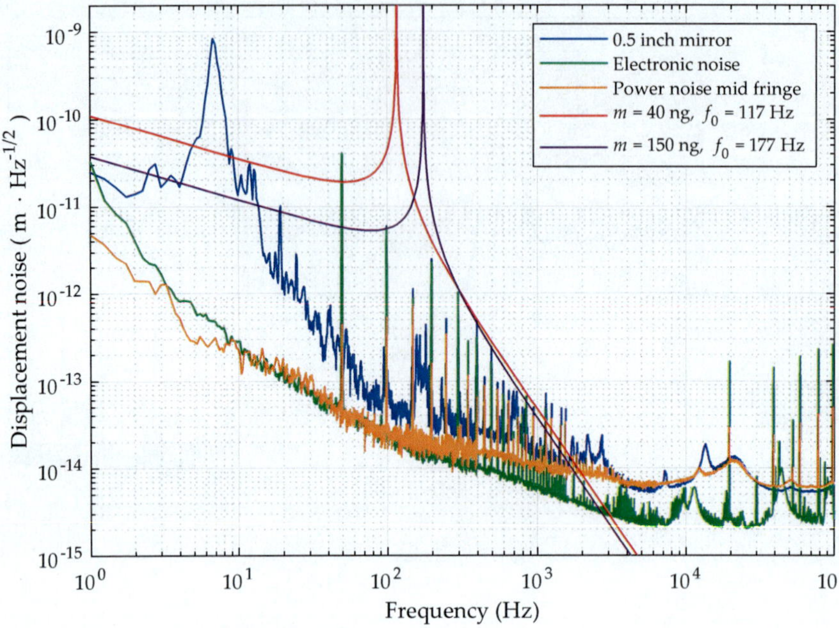

Fig. 5.10 Final interferometer sensitivity achieved with the 0.5 inch mirror (blue curve) compared with the expected structural thermal noise displacement for two micro-oscillators, calculated at room temperature and with a quality factor of 10^4 (red and purple curves). A projection of the total electronic noise (green curve), and the sensing beam power noise coupling via the mid-fringe readout (yellow curve) is shown as a reference

this experiment the following changes should be possibly implemented: a dark fringe lock to avoid the direct power noise coupling at the interferometer output, further reduction of electronic noise in the MI control loop, and higher vibration isolation.

5.2 Sensing Noise: Interferometer Sensitivity with Micro-oscillators

After optimizing the interferometer sensitivity with the 0.5 inch mirror, the chip containing the micro-oscillators was placed in the translation stage and the interferometer displacement noise was measured with different oscillators.

5.2.1 Displacement Noise with Micro-oscillator LA7

The first measurements were performed with the micro-oscillator LA7,[1] which was designed to have a fundamental resonance frequency between 150 and 200 Hz, and a mass between 100 and 20 ng. This oscillator had a mirror pad with a radius of 50 μm, and a cantilever with length of 200 μm and width of 10 μm.

From the measurements performed with the 0.5 inch mirror, it is expected that the oscillator's thermal noise will dominate the interferometer displacement noise at the mid-frequency regime of this experiment. As discussed in Sect. 3.2, the contribution of viscous damping to thermal noise displacement is reduced by decreasing the pressure inside the vacuum chamber. This effect can be observed in Fig. 5.11, which shows the interferometer displacement noise at different pressures (solid line curves). A fit of the total thermal noise displacement of the micro-oscillator for each pressure is displayed by the dashed curves. The fit was calculated by substituting the total loss angle $\phi = \phi_v + \phi_s$ from the viscous and structural damping contributions into Eq. 3.67. The values for the viscous quality factor Q_v were adjusted for each pressure, while the structural quality factor Q_s was assumed to be the same for all pressures. As a result, a value of $Q_s = 1.7 \times 10^4$ was obtained, which is within the expected range for this generation of micro-oscillators at a room temperature [73]. The measurements show that the contribution of structural damping is significant at pressures below 6×10^{-3} mbar. The maximum viscous quality factor of $Q_v = 10^5$ was achieved at a pressure of 6×10^{-5} mbar. This value is two times smaller than what is predicted from Eq. 3.71, where a viscous damping model taking into account the thickness of the mirror pad and inelastic collisions with water molecules was considered. This model seems to be more appropriate to describe the conditions in this experiment than the model proposed in Eq. 3.69, which considers only elastic collisions with residual nitrogen and predicts a factor of $Q_v = 5 \times 10^6$. The purple curve shows the expected thermal noise displacement calculated solely with a contribution from structural damping. By comparing it with the blue dashed curve, it is possible to conclude that the residual viscous damping still has a non negligible contribution to thermal noise. At the fundamental resonance frequency, the contribution to the total loss angle is 14.5%, while at higher frequencies this contribution is increased since $\phi_v = \Omega/(\Omega_0 Q_v)$. To reach the regime displayed by the purple curve, a viscous quality factor of at least $Q_v = 10^7$ is needed. In the future this could be achieved by heating the vacuum chamber in order to reduce the number of water molecules. Furthermore, it was observed that the displacement noise at 6×10^{-5} mbar was not considerably improved in comparison with a pressure of 10^{-4} mbar (not shown in the plot), which was preferred since the pressure increase over a typical set of measurements was less significant at this value.

The longitudinal (177 Hz), yaw (1296 Hz), and pitch (3328 Hz) fundamental resonance frequencies for this oscillator were determined from the peaks displayed

[1] The micro-oscillators are named according to their position in the chip, following the schematics presented in Fig. 4.2b. In this case, L stands for the left block, A stands for the first row, and 7 stands for the 7th oscillator from left to right in the schematics.

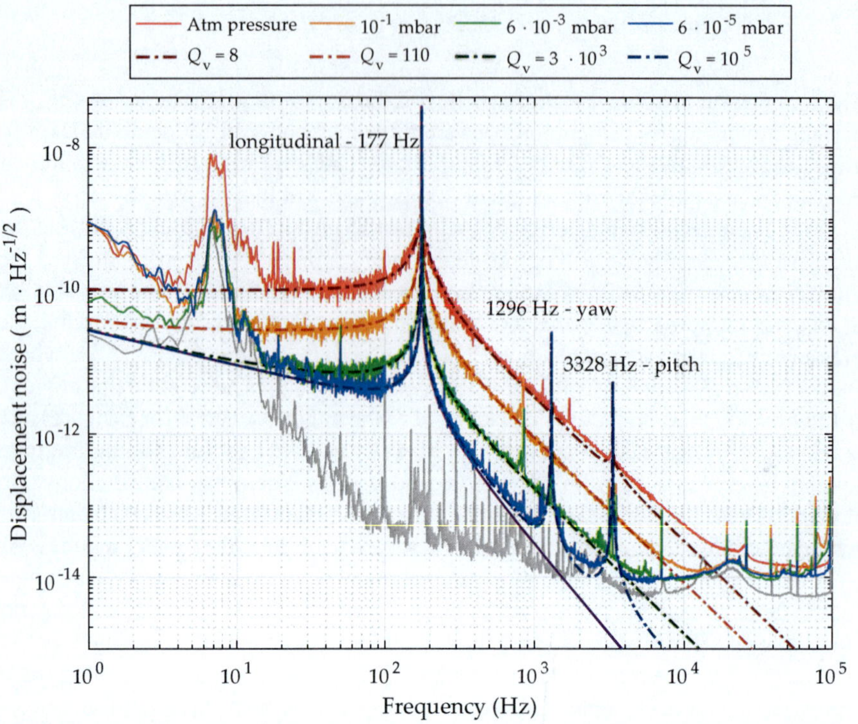

Fig. 5.11 ASD of the interferometer displacement noise measured with the micro-oscillator LA7 for different pressures in the vacuum chamber (solid lines). A thermal noise fit for each measurement is displayed by the corresponding dashed lines. The purple curve shows the expected structural thermal noise displacement at room temperature for a $Q_s = 1.7 \times 10^4$. For comparison, the gray curve shows the interferometer displacement noise measured with the 0.5 inch mirror. The thermal noise curves were plotted with the parameters from Table 5.1. For all measurements the transfer beam was turned on, with a mean power of 46 mW

in the plot. In addition to that, all theoretical curves were plotted considering a mass of 150 ng, which was obtained from the radiation pressure transfer function of this micro-oscillator (described in Sect. 5.3).

An interesting feature in the displacement noise which was observed with all micro-oscillators is related to the 7 Hz broad peak, which is at the resonance frequency of the VIB 100 isolation feet. The height of the peak is approximately one order of magnitude higher in atmospheric pressure than in vacuum, were the height remained the same for all pressures and was equal to the measurement with the 0.5 inch mirror. This additional noise in air is most likely caused by the relative motion between the breadboard and the micro-oscillator's mirror pad, whose motion is now damped by the air. This relative motion causes a distance change between the mirror pad and the interferometer beamsplitter (which is expected to move rigidly with the breadboard) and will couple as displacement noise. This effect seems to be in phase with the coupling path of the noise at 7 Hz measured in vacuum (which was not fully understood), resulting in an additional displacement noise in air.

It is important to note that, in order to preserve the micro-oscillator's alignment, the transfer beam was kept on during the measurements with a mean power of 46 mW. The radiation pressure displacement caused by the transfer beam power noise in this case is non negligible. This is the reason why the thermal noise fits displayed by the dashed blue and green curves are slightly below the measured displacement noise at frequencies between 20 Hz and 2 kHz. To account for this effect, the corresponding quality factors for these curves were obtained by fitting the displacement noise with an uncorrelated sum of the radiation pressure and the thermal noise displacements. Furthermore, the transfer beam power noise was the source of the observed excess displacement noise at frequencies below 5 Hz. Above 2 kHz the displacement noise was limited by technical noise sources, as indicated by the gray curve showing the displacement noise measured with the 0.5 inch mirror. In conclusion, the displacement noise was dominated by thermal noise for frequencies ranging from 20 Hz until several kHz.

5.2.2 *Displacement Noise with Micro-oscillator LC6*

One of the most promising oscillators available in this experiment is the LC6, since it has the highest transfer coefficient τ (defined in Eq. 2.3). This oscillator was designed to have a mass of 40 ng and a resonance frequency smaller than 117 Hz, which is very close to the parameters used to calculate the fundamental limits of this scheme in Chap. 3. To achieve such values, the cantilever was designed to have a length of 285 μm, which is the longest in this chip, and a width of 6 μm. In addition to that, the mirror pad was designed to have a radius of 25 μm, but, due to etching problems during the manufacturing process, the size of the coating layers was not homogeneous and it was smaller than its design value (see Fig. 4.3). Hence, it is expected that the oscillator will have a smaller mass than its design value, which will consequently shift the longitudinal resonance frequency to a higher value.

Due to its small size, this oscillator was first aligned to the sensing and transfer beams with atmospheric pressure in the chamber. A fine alignment was performed by tilting the sensing and transfer beams individually to the oscillator, to make sure both beams are well centered to the mirror pad, as described in Sect. 4.3. The pitch and the yaw resonance peaks were reduced until there were not visible in the displacement noise, and the interferometer achieved a maximum contrast of 93%. It is worth to note that the support layer of the mirror pad was facing the sensing beam, which most probably prevented scattering or beam deformations arising from inhomogeneities in the upper coating layers to potentially reduce the contrast.

Figure 5.12 shows the interferometer displacement noise measured with the LC6 oscillator at atmospheric pressure and at 10^{-2} mbar. The longitudinal resonance frequency of this oscillator was identified to be at 219 Hz. The maximum thermal noise displacement in air occurred at a slightly lower frequency, due to the

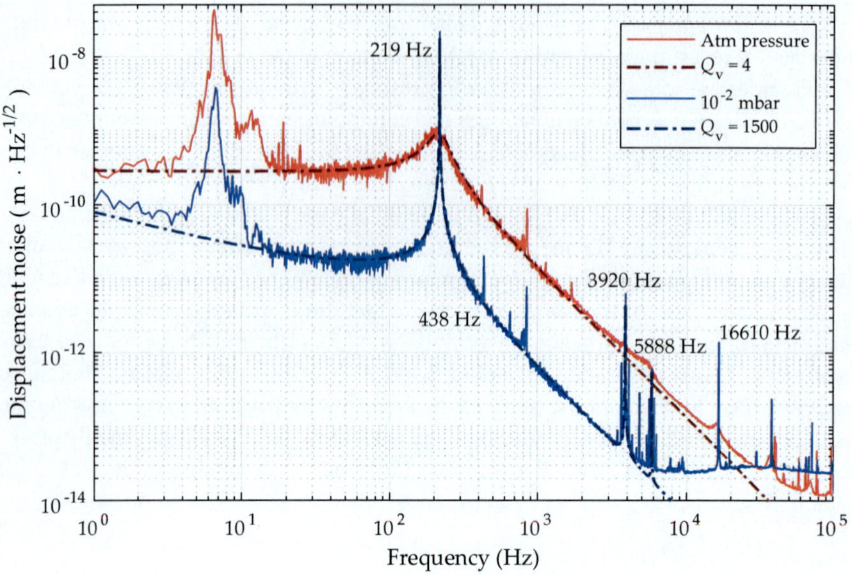

Fig. 5.12 ASD of the interferometer displacement noise measured with the micro-oscillator LC6 for different pressures (solid lines) in the vacuum chamber. A thermal noise fit for each measurement is displayed by the corresponding dashed lines. The thermal noise estimations were made with the parameters of Table 5.1. The transfer beam was turned off during the measurements

high viscous damping contribution. The figure also shows the fundamental yaw (3920 Hz), pitch (5888 Hz), and side-to-side (16610 Hz) resonance peaks. The dashed curves shows a fit of the total thermal noise displacement, calculated with the same structural quality factor obtained from the LA7 oscillator ($Q_s = 1.7 \times 10^4$), and considering a mass of 25 ng. These values would lead to a transfer coefficient of $\tau = 1.6 \times 10^3$ rad · W^{-1} for frequencies below the longitudinal resonance.

The experiment with the LC6 oscillator was much more challenging than with other oscillators. The first challenge was to achieve a stable and long lasting lock of the interferometer to the mid-fringe. The amplitude of the oscillator's motion at the longitudinal resonance frequency caused a particularly large oscillation in the error signal of the interferometer control loop. These oscillations increased, in a timescale of minutes, to values that were large enough to bring the interferometer out of the linear range around the mid-fringe operational point. A similar behavior was observed with the LA7 oscillator, but only at pressures below 10^{-4} mbar, and with a smaller amplitude in the error signal. In that case, a reduction in the open loop gain of the interferometer provided a better stability. A possible explanation for this effect is that the PZT mirror motion, which follows the oscillator displacement, end up exciting the oscillator's motion at its longitudinal resonance frequency, resulting in a positive feedback situation that brings the interferometer out of lock. The effect with the LC6 oscillator, however, could not be improved by reducing, neither by increasing, the interferometer control loop gain. In a future experiment with similar oscillators, a notch filter centered at the longitudinal resonance frequency should be implemented in the MI control loop as an attempt to improve the locking conditions.[2] The difficulty on locking the interferometer prevented improvement to the alignment of the oscillator in vacuum. In addition to that, it was not possible to lock the interferometer with pressures lower than 10^{-2} mbar due to the higher displacement of the oscillator at the longitudinal resonance frequency.

Another challenge with this oscillator was related to the increase in the transfer beam power, which was realized in vacuum. By increasing the transfer beam power up to 1 mW, a longitudinal static displacement of 1.4×10^{-7} μm is expected for this oscillator. Since this displacement is not purely longitudinal, the mirror pad needs to be readjusted in order to restore the interferometer contrast (details described in Sect. 5.4.2). During this realignment, the micro-oscillator broke. It was observed in the CCD cameras that the beams were, for a short time, completely misaligned to the mirror pad during the step motions of the translation stage, which excites the oscillator motion. Hence, the damage was most probably caused by the mechanical stress in the oscillator, induced by radiation pressure from the sensing and transfer beams during the step motion. An alternative procedure for a future experiment with a similar oscillator, would be to increase the transfer beam power and restore the mirror pad alignment with atmospheric pressure in the chamber to reduce the oscillator motion.

Because of the damage, no other measurements were made with this oscillator.

5.2.3 Displacement Noise with Micro-oscillator LC5

The last micro-oscillator implemented in this experiment was the LC5, which had a mirror pad radius of 60 μm, and a cantilever with length of 200 μm and width of 12 μm (see Fig. 4.3). The interferometer displacement noise measured with this oscillator is shown in Fig. 5.13, together with the previous measurements with the oscillators LA7, LC6, and with the 0.5 inch mirror. The green curve shows that the LC5 oscillator had a fundamental longitudinal resonance frequency of 151 Hz. A large displacement noise can be observed at 150 Hz, which is an harmonic of the utility frequency of 50 Hz, and is very close to oscillator's longitudinal resonance frequency. This displacement was caused by the power noise of the transfer beam, which had a mean power of approximately 50 mW in this measurement. A thermal noise fit, similar to what described in Sect. 5.2.1, resulted in a structural quality factor of $Q_s = 6 \times 10^3$, and a viscous quality factor of $Q_v = 10^5$ at 10^{-4} mbar for this oscillator.

[2] A notch filter was not implemented in this setup due to the damage in this oscillator after turning the transfer beam on.

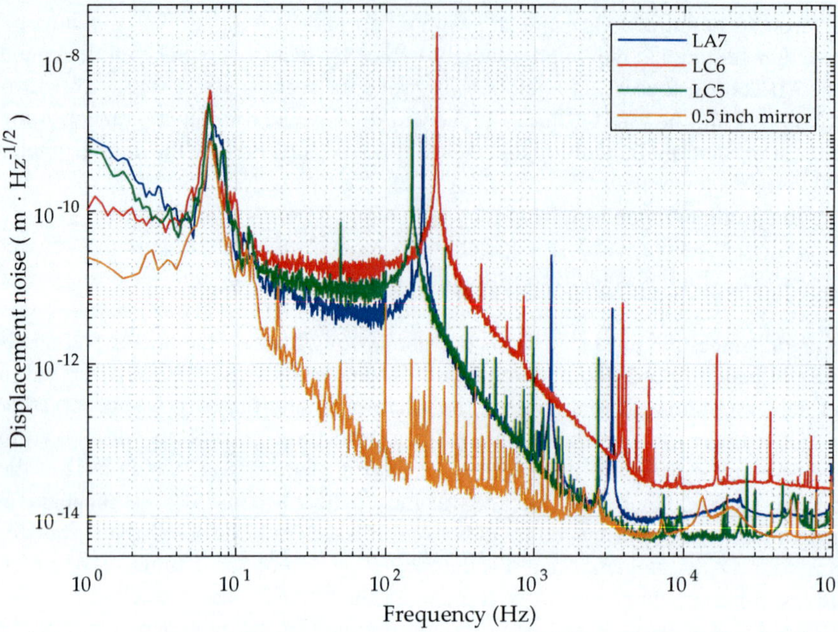

Fig. 5.13 ASD of the interferometer displacement noise measured with the 0.5 inch mirror, and with the micro-oscillators LA7, LC6, and LC5. The pressure in the chamber was approximately 10^{-4} mbar for the LA7 and LC5 oscillators, 10^{-2} mbar for the LC6 oscillator, and atmospheric pressure for the 0.5 inch mirror. The transfer beam was turned on, with a power of approximately 50 mW, in the measurements with the LA7 and LC5 oscillators

5.3 Power Sensing

After the sensing noise characterization, the interferometer was used as an in-loop sensor for power fluctuations/modulations of the transfer beam. First, the radiation pressure transfer function R, from transfer beam power modulations to micro-oscillator displacement, was measured. The results are shown in Fig. 5.14 for the LA7 and LC5 oscillators (full curves). The dashed curves display the corresponding transfer function fit, çalculated with Eq. 2.26.

The effective mass of the longitudinal mechanical mode was obtained from the fit, which resulted in a mass of 150 ng for the LA7, and a mass of 190 ng for the LC5 oscillator. The parameters for the oscillators used in this thesis are summarized in Table 5.1. The radiation pressure transfer function was also measured for different pressures inside the chamber and, as expected, no significant differences were observed apart from the total loss angle.

In a second step, a comparison between measurements of the free running power fluctuations of the transfer beam sensed by the interferometer (in-loop measurement) and by the OOL PD was made for the LA7 oscillator. The projection $\mathrm{RPN_{il}}$ from the interferometer displacement readout x to transfer beam relative power noise was calculated with the following equation:

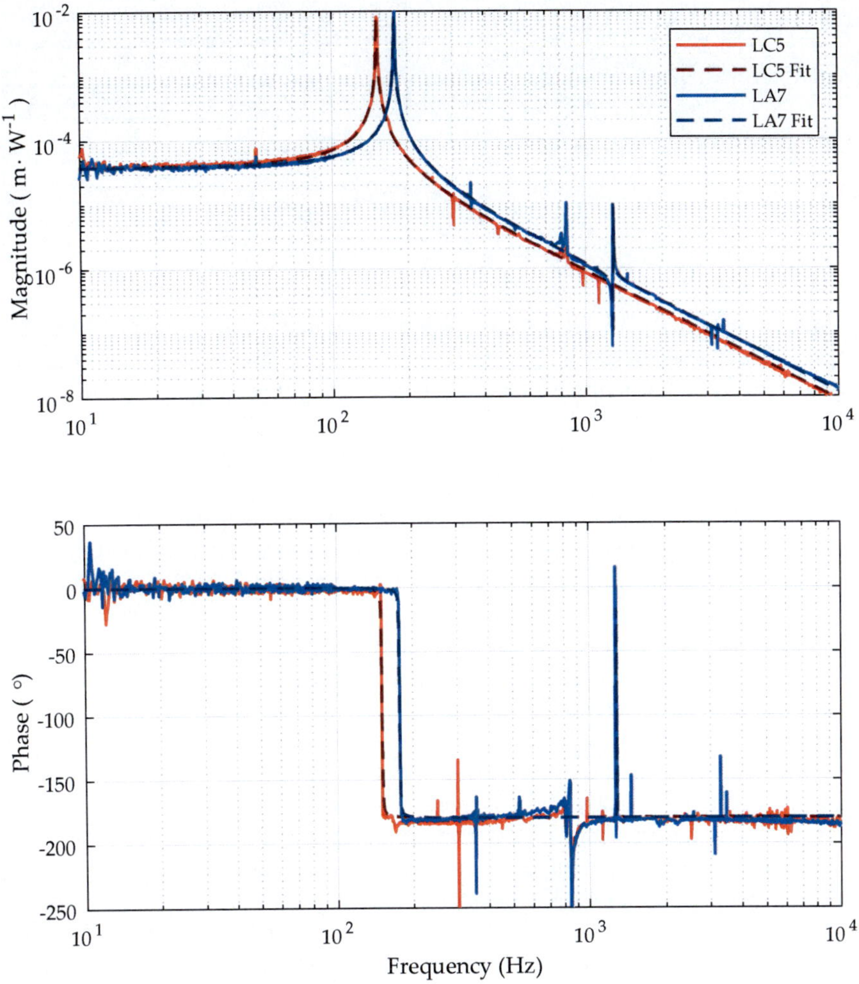

Fig. 5.14 Bode plot of the radiation pressure transfer function R, measured with the micro-oscillators LA7 and LC5. The dashed curves are a fit of the transfer function, calculated according to Eq. 2.26, and considering the parameters displayed from Table 5.1. The measurements were performed with pressures around 10^{-4} mbar in the vacuum chamber, and with 46 mW of power in the transfer beam

$$\mathrm{RPN_{il}} = \left| \frac{x}{R \cdot \overline{P}_{\mathrm{t}}} \right| = \left| \frac{V_{\mathrm{s}}\,(1 + G_{\mathrm{MI}})}{HR \cdot \overline{P}_{\mathrm{t}}} \right| . \tag{5.10}$$

The mean power on the transfer beam was initially set to 46 mW, which should be far below the damage threshold of this oscillator. Apart from frequencies below 5 Hz, the absolute power fluctuations of the transfer beam for 46 mW are too small to drive the oscillator's motion above the interferometer displacement noise. This resulted in

Table 5.1 Parameters for the micro-oscillators LA7, LC6, and LC5: mirror pad radius, cantilever length and width, effective mass m of the longitudinal mechanical mode, fundamental longitudinal resonance frequency f_0, and transfer coefficient τ

	Pad (μm)	Length (μm)	Width (μm)	m (ng)	f_0 (Hz)	τ $(\text{rad} \cdot \text{W}^{-1})$
LA7	50	200	10	150	177	424
LC6	<25	285	6	~25	219	1660
LC5	60	200	12	190	151	460

a discrepancy between the in and out-of-loop power noise measurements, as shown by the red and blue curves in Fig. 5.15. Hence, to demonstrate a power sensing and power stabilization at low powers, a broadband white noise was imprinted on the transfer beam. As shown by the green and yellow curves, the imprinted noise was high enough such that the in and out-of-loop power noise measurements overlap for frequencies between 10 Hz and 5 kHz. At higher frequencies, the in-loop measurement is still dominated by sensing noise, which is approximately flat in this frequency regime, and therefore the in-loop projection follows a shape of approximately f^2, as expected from the radiation pressure transfer function in Eq. 5.10. An in-loop power measurement without imprinting white noise should be possible at frequencies up to the kHz regime for a transfer beam power higher than 200 mW.

Figure 5.16 shows the interferometer displacement, measured with the LA7 oscillator, when no noise is applied to the transfer beam (green curve), and when a white noise with amplitude of 10^{-6} W $\cdot$ Hz$^{-1/2}$ and 5×10^{-7} W $\cdot$ Hz$^{-1/2}$ is imprinted on the transfer beam (blue and red curves). As expected, the green curve follows the thermal noise fit of the oscillator (dashed green curve), while the blue and the red curves follow the radiation pressure displacement driven by the transfer beam (dashed blue and red curves). In addition to that, the increase in the radiation pressure displacement was linear with the applied white noise.

In conclusion, the measurements demonstrated that the interferometer can be used as a power sensor for the transfer beam for frequencies up to 6 kHz, in agreement with what expected from the sensing noise characterization.

5.4 Power Stabilization via Radiation Pressure

After demonstrating the feasibility and the limitations of using the interferometer as a power sensor for the transfer beam, the power stabilization via radiation pressure control loop was enabled. This section presents the out-of-loop beam power noise performance from this proof-of-principle experiment for the micro-oscillators LA7 and LC5.

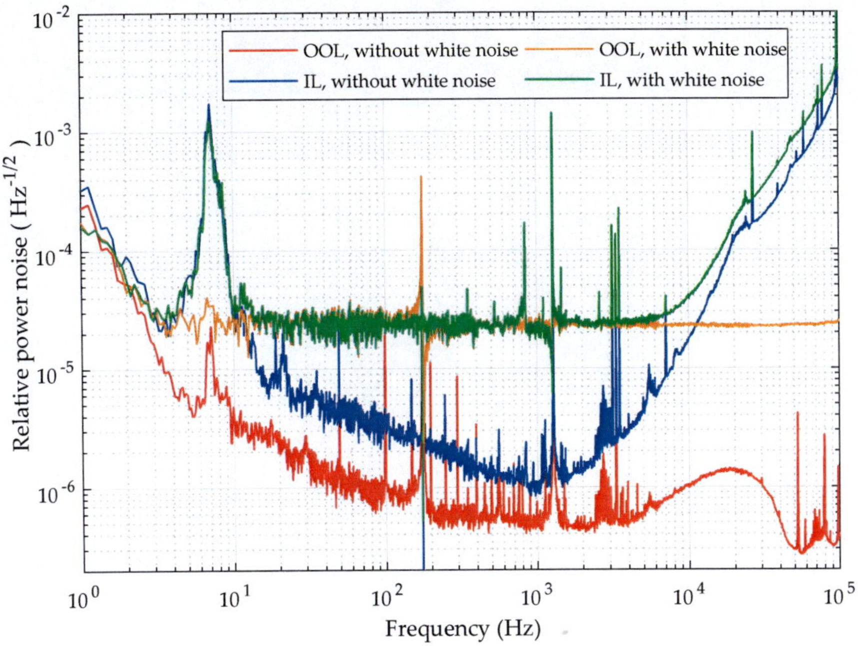

Fig. 5.15 ASD of the out-of-loop (OOL) and in-loop (IL) measurements of the free running relative power noise of the transfer beam before (red and blue curves), and after (green and orange curves) imprinting a broadband white noise in the transfer beam. The measurements were performed with the LA7 micro-oscillator and with a pressure of 2×10^{-4} mbar in the vacuum chamber. The transfer beam power was 46 mW

5.4.1 Stabilization Control Loop

A block diagram of the power stabilization control loop is shown in Fig. 5.17. The absolute free running power fluctuations of the transfer beam δP_{fr} are detected by the in-loop sensor, formed by the Michelson interferometer and the micro-oscillator, which has a complex transfer function $S = R \cdot H_1 \cdot H_2$. The sensor signal V_{s} at the output of the MI control loop is amplified by a controller with a complex transfer function C_{RP}, and the resulting control signal V_{c} is sent to the AOM, which has a complex transfer function A_{RP}. The residual power fluctuations in the stabilized out-of-loop beam are represented by δP_{ool}. This experiment had the aim to suppress power fluctuations at frequencies above 1 Hz. If reference tracking is needed, to avoid power drifts for example, then a reference signal should be subtracted from the sensor signal in a similar fashion as in the MI control loop.

The controller consisted on four stages: a calibration stage, a differentiator stage, an integrator stage, and a proportional gain stage. Figure 5.18 presents a measurement of the transfer function of the individual stages, as well as the total controller transfer function and its fit (black dashed curve). The calibration stage (red

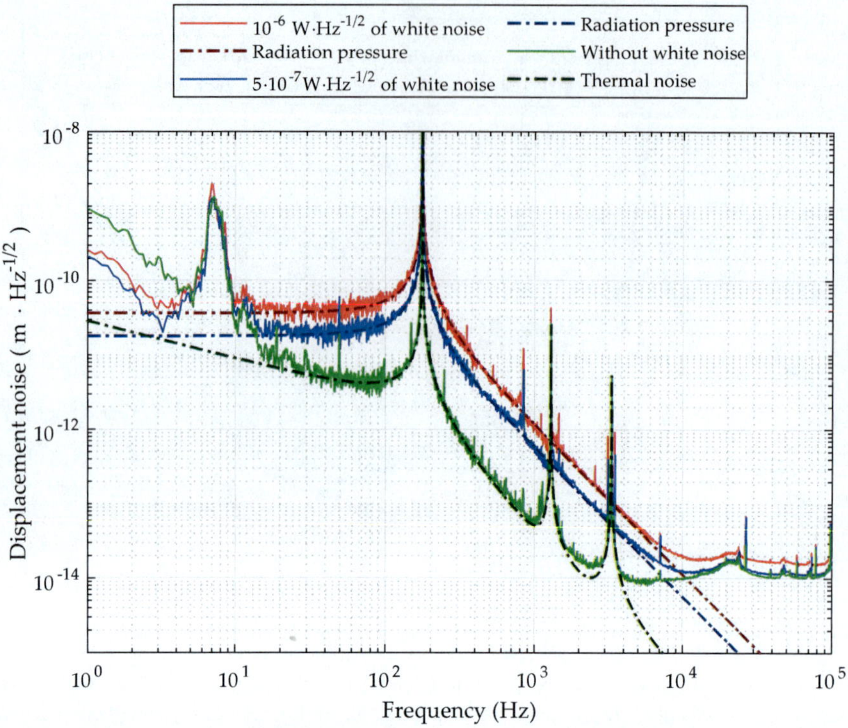

Fig. 5.16 ASD of the interferometer displacement with (blue and red curves) and without (green curve) white noise imprinted on the transfer beam. The dashed green curve represents the thermal noise displacement of the oscillator, and the dashed red and blue curves represent the expected radiation pressure displacement driven by a frequency independent transfer beam power noise corresponding to the applied white noise. The measurements were performed with the LA7 micro-oscillator and with a pressure of 10^{-4} mbar in the vacuum chamber. The transfer beam power was 46 mW in all measurements

curve) had the purpose of providing gain at low frequencies in order to compensate for the suppression in the sensor signal by the MI control loop gain. Its transfer function had an approximate shape of $1 + G_{\mathrm{MI}}/f$ for frequencies between 1 and 410 Hz. This stage also contains a high pass filter with a corner frequency of 3 Hz to suppress possible DC offsets, that could alter the transfer beam mean power. An additional frequency dependent gain at low frequencies was provided by the integrator (blue curve), which had a lower corner frequency of approximately 8 Hz. The differentiator had the purpose of compensating the 180° phase loss after the oscillator's longitudinal fundamental resonance frequency, and it added a maximum positive phase of approximately 100° at 2 kHz. At higher frequencies, a low pass filter stage with a corner frequency of 25 kHz was added to the differentiator to suppress high-frequency resonances of the operational amplifiers used in this stage, and also to reduce contributions caused by ground loops at

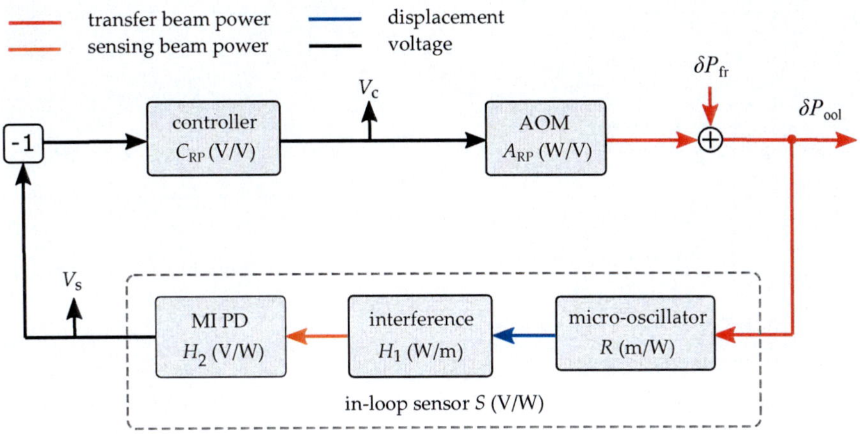

Fig. 5.17 Block diagram of the power stabilization via radiation pressure control loop. The absolute free running power fluctuations δP_{fr} of the transfer beam are sensed by the interferometer containing the micro-oscillator, which has a total complex transfer function S. The sensor signal V_{s} is amplified by a controller with transfer function C_{RP}, and the resulting control signal V_{c} is sent to the AOM (power actuator), which has a transfer function A_{RP}. The residual power fluctuations in the stabilized out-of-loop beam are represented by δP_{ool}

$\approx$ 200 kHz, which would cause instabilities in the control loop. The AOM had a flat transfer function up to the MHz regime.

Figure 5.19 shows the open loop transfer function G_{RP} measurement performed with the micro-oscillator LA7. The dashed black curve represents the fit of the transfer function, calculated with the following equation:

$$G_{\mathrm{RP}} = \frac{S\, C_{\mathrm{RP}}\, A_{\mathrm{RP}}}{(1 + G_{\mathrm{MI}})} \,. \tag{5.11}$$

At high frequencies the magnitude deviated from the fit since the injected excitation at the AOM was not sufficient to overcome the interferometer displacement noise. The maximum UGF achieved with this configuration was around 7 kHz with a phase margin of 65°. The measurement shows that, in principle, the UGF could be increased to values up to 20 kHz with a phase margin of 30°. However, such a value was not possible to achieve in practice due to peaks at frequencies around 200 kHz which would have an amplitude larger than 0 dB and a phase loss close to 180° in the transfer function measurement. The origin of these peaks was not well understood, but measurements showed that a reduction on the peaks was possible by changing the electronic supply connections in the experiment, which is evidence for a cause connected to ground loops. The additional phase loss between the measurement and the fit curve at frequencies above 5 kHz is due to phase delays introduced by the AOM, and by the operational amplifiers used in the controller.

The horizontal alignment of the micro-oscillator was not optimal in this measurement, which explains the coupling from the yaw motion of the oscillator to the trans-

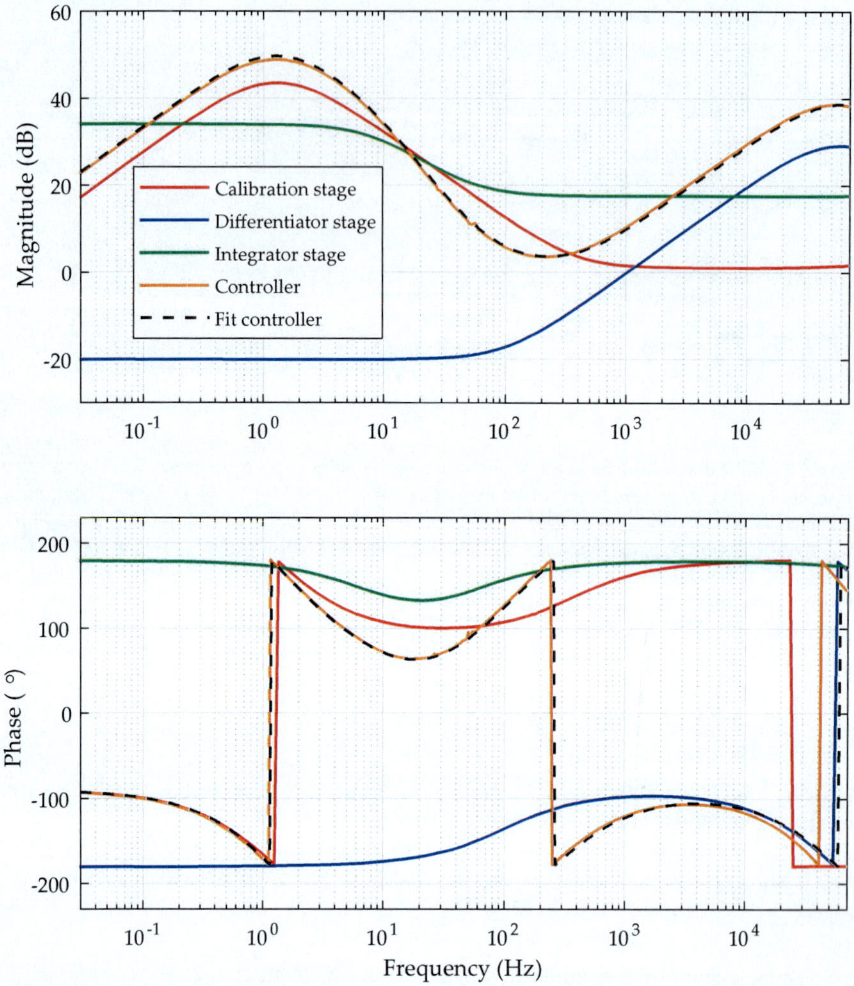

Fig. 5.18 Bode plot of the measured transfer functions of the individual stages of the controller, and of the total controller transfer function (yellow curve) and its fit (dashed black curve)

fer function. The fact that the corresponding magnitude from this resonance crosses 0 dB is not an issue for the stability of the control loop since it has a corresponding positive phase. For resonances with a high amplitude, however, nonlinear effects can take place in the control loop and bring it to an unstable regime. Therefore, before turning the power stabilization control loop on, the oscillator was aligned to reduce the coupling at the yaw and pitch resonance frequencies to as small as possible.

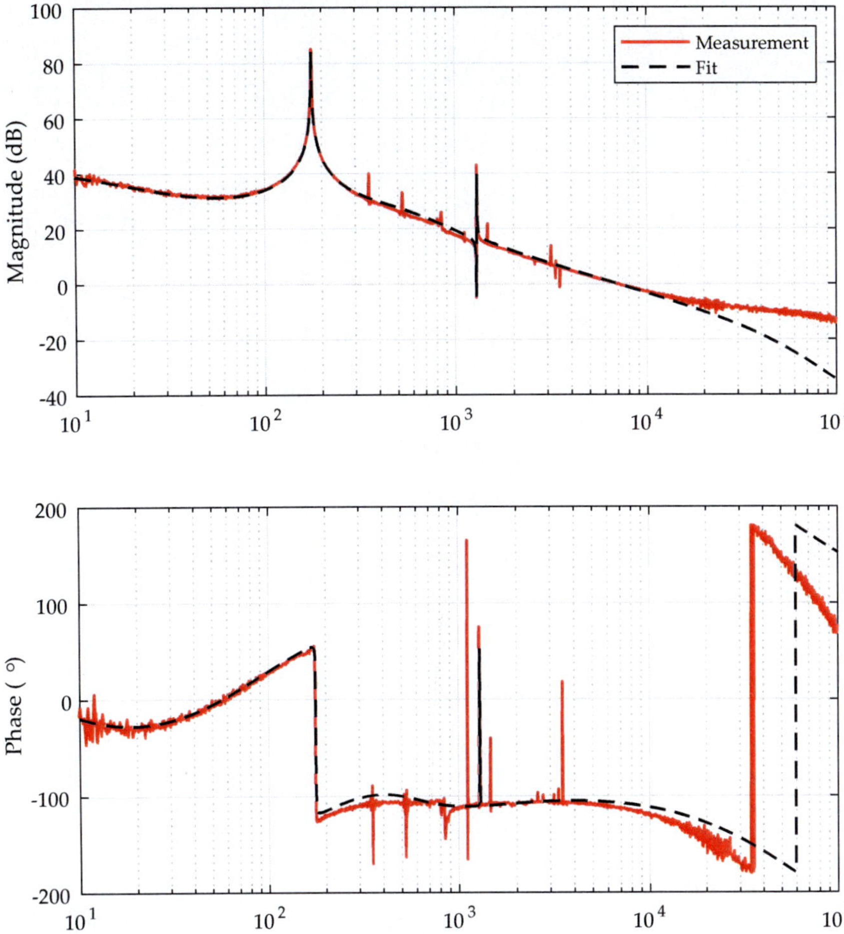

Fig. 5.19 Bode plot of the power stabilization open-loop transfer function measurement (red curve), and its fit (dashed black curve). The measurement was performed with the LA7 micro-oscillator, and with a pressure of 10^{-4} mbar in the vacuum chamber

5.4.2 Power Stabilization with Micro-oscillator LA7

In a first step, the stabilization was performed at different pressures in the vacuum chamber and with a fixed transfer beam power of 46 mW. The results are shown in Fig. 5.20, where the RPN measured by the OOL PD is plotted when the power stabilization control loop is turned off, and turned on with different pressures in the chamber. The corresponding dashed curves represent the expected power noise performance, calculated as an uncorrelated sum of the micro-oscillator's thermal noise displacement projected to the out-of-loop beam, and the expected free running

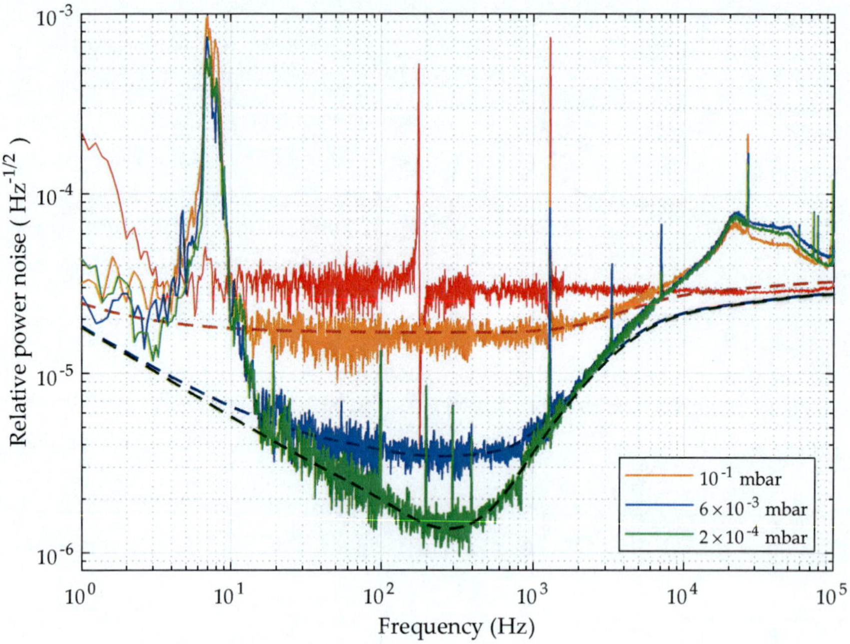

Fig. 5.20 ASD of the relative power noise measured by the OOL PD when the power stabilization control loop is turned off (red curve), and when the loop is turned on with different pressures in the vacuum chamber (orange, blue, and green curves). The dashed curves represent the expected power noise performance and were calculated as an uncorrelated sum of the micro-oscillator's thermal noise, and the expected free running noise reduction by the control loop. The measurements were made with a transfer beam power of 46 mW and with an imprinted broadband white noise. The power stabilization open loop UGF was 7 kHz for all measurements

noise reduction (of a frequency independent RPN $= 2.8 \times 10^{-5}$ $\text{Hz}^{-1/2}$) by the open loop gain G_{RP}. The plot shows that the power stability was increased by lowering the pressure, which is expected from the reduction of the viscous thermal noise contribution to the sensing noise. At frequencies between 15 and 850 Hz, the free running noise was suppressed by a frequency independent factor of approximately 1.8 at a pressure of 10^{-1} mbar, which agrees with the viscous thermal noise limit. At a pressure of 2×10^{-4} mbar, the relative power noise followed a $1/f$ slope from 15 Hz until 300 Hz, where the lowest value of 1.3×10^{-6} $\text{Hz}^{-1/2}$ was achieved, in agreement with a structural thermal noise limitation. At frequencies higher than 340 Hz, the performance was limited by the control loop gain. The peak around 7 Hz was imprinted on the out-of-loop beam with approximately the same height for all pressures. A curious feature observed in the free running power noise measured with the OOL with all oscillators, also when no white noise was applied to the transfer beam, was a resonance behavior at the longitudinal resonance frequency. It is suspected that this effect was caused by the interference between an apparent

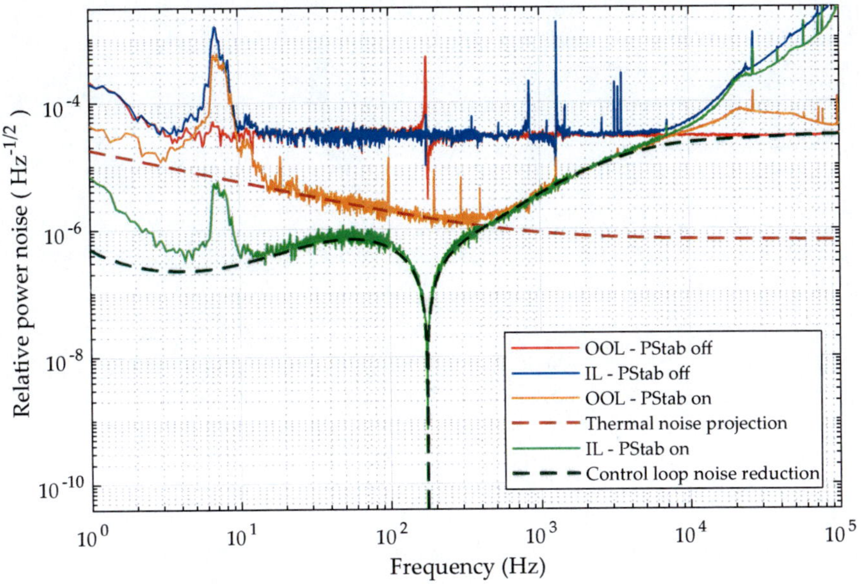

Fig. 5.21 ASD of the transfer beam relative power noise obtained from the out-of-loop (OOL) and in-loop (IL) measurements when the power stabilization control loop is turned off (red and blue curves), and on (orange and green full curves). The dashed orange curve shows the micro-oscillator total thermal noise displacement projected to the out-of-loop beam, and the dashed green curve shows the expected control loop power noise reduction. The measurements were performed with a transfer beam power of 46 mW, with a pressure of 2×10^{-4} mbar in the vacuum chamber, and with an open loop UGF of 7 kHz

power noise due to beam jitter at the photodiode and the free running power noise in the transfer beam.

A comparison between the out-of-loop relative power noise measured by the in-loop and out-of-loop sensors is presented in Fig. 5.21. The dashed orange curve depicts the micro-oscillator total thermal noise contribution, calculated from the fit presented in Sect. 5.2.1, projected to the out-of-loop beam. The projection overlaps with the out-of-loop power noise measurement for frequencies between 15 and 300 Hz, and below 4 Hz, confirming that the power stability is thermal noise limited in this frequency regime. The dashed green curve depicts the expected free running power noise reduction by the control loop, considering an initial frequency independent relative power noise of 2.8×10^{-5} Hz$^{-1/2}$. This curve agrees with the performance of the in-loop detector from 10 Hz to 5 kHz, where a minimum value of 1.4×10^{-11} Hz$^{-1/2}$ was achieved at the oscillator's longitudinal resonance frequency. The additional noise in the in-loop measurement at low frequencies with respect to the noise reduction curve was because the free running RPN_{ool} at those frequencies was higher than the one used in the calculation. At high frequencies, the projected interferometer displacement noise is higher than the free running RPN_{ool}, and therefore the in-loop measurement does not agree with the noise reduction curve. The

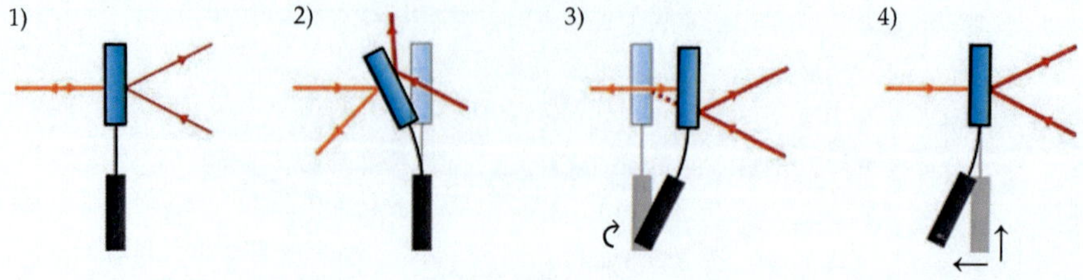

Fig. 5.22 Illustration of the procedure for the micro-oscillator realignment after increasing the transfer beam power

overlap between the noise reduction curve and the out-of-loop measurement above 450 Hz confirms that the out-of-loop stability is limited by the open loop gain in this frequency regime. Furthermore, the thermal noise projection indicates that increasing the open loop UGF could result in a relative power stability of 7×10^{-7} Hz$^{-1/2}$ at the kHz regime for a transfer beam power of 46 mW.

In sequence, the out-of-loop power stability was measured for different transfer beam mean powers. The power was gradually increased over steps around 10 mW, and in each step the micro-oscillator's degrees of freedom had to be readjusted in order to compensate for the misalignment caused by the transfer beam. As shown in Fig. 5.22, an increase in the transfer beam power induces a static longitudinal displacement plus a pitch rotation in the oscillator (step 2). This rotation was enough to change the interferometer contrast,[3] and also the position of the mirror pad center with respect to the sensing and transfer beams. To restore the interferometer contrast, the oscillator chip was rotated in the pitch direction, as illustrated in step 3. After this rotation, however, the mirror pad is longitudinally displaced from its initial position and it is no longer centered to the transfer beam. Hence, the chip needed to be horizontally and vertically displaced in order to restore the initial alignment. Steps 3 and 4 had to be repeated over a few times to find the optimum position, since the mirror pad also has a small rotation (not shown in the figure) in step 3 caused by the misalignment with respect to the transfer beam.

The out-of-loop power stability performance for a transfer beam power of 46 and 119 mW are shown in Fig. 5.23. The dashed curves were calculated as an uncorrelated sum of the micro-oscillator's thermal noise and the expected free running noise reduction by the control loop. The reduction in the relative power noise after an increase in the transfer beam power can be observed at frequencies below 500 Hz. A minimum relative power noise of 6.8×10^{-7} Hz$^{-1/2}$ at a frequency of 200 Hz was achieved for a transfer beam power of 119 mW. It is worth to note that the UGF for the measurement at 119 mW was 8 kHz, which was slightly higher than for the measurement at 46 mW (7 kHz). This is the reason why the power noise at high frequencies is not the same for the different powers, which should be the case if the UGF was equal in both measurements, since the performance is limited by the open loop gain. The purple curve represents the minimum out-of-loop rela-

[3] A power change of 13.8 mW changed the contrast from 97 to 93%, for example.

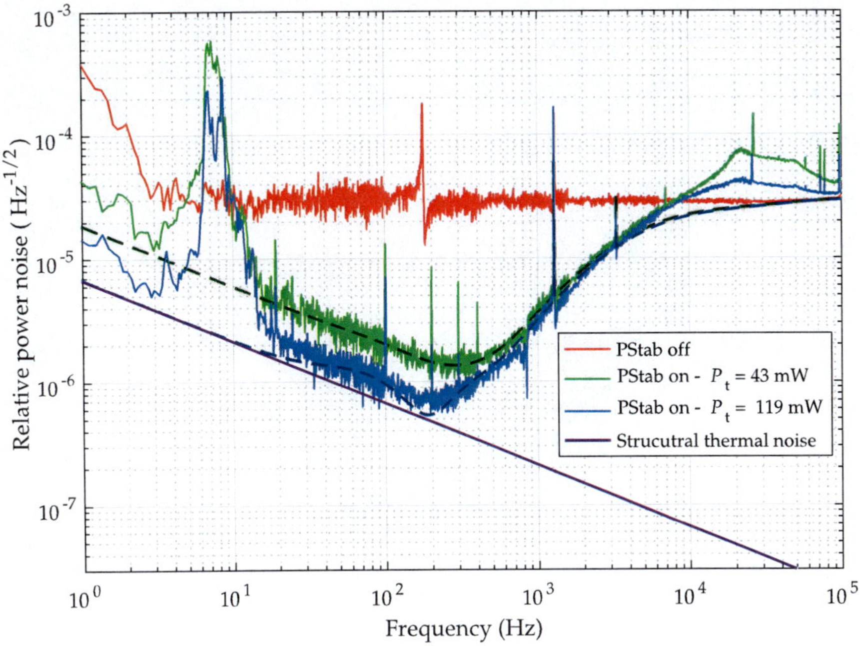

Fig. 5.23 ASD of the relative power noise measured by the out-of-loop sensor when the power stabilization control loop is turned off (red) and when the loop is on for different transfer beam powers. For all measurements a broadband white noise was imprinted at the transfer beam. The measurements were performed with a pressure of 10^{-4} mbar at the vacuum chamber, and with an open loop UGF of 7 kHz for a mean power of 46 mW, and of 8 kHz for a mean power of 119 mW. The purple curve represents the structural thermal noise projection for this micro-oscillator, considering a transfer beam power of 119 mW

tive power noise that could, in principle, be achieved for a transfer beam power of 119 mW, if the performance was solely limited by the structural thermal noise of the micro-oscillator.

Higher stability levels could be achieved by further increasing the transfer beam mean power. However, the micro-oscillator broke at a value of approximately 150 mW. The breaking mechanism was not entirely understood.

5.4.3 Power Stabilization with Micro-oscillator LC5

The last power stabilization experiment was realized with the LC5 oscillator. In this part of the experiment, the controller of the power stabilization loop was modified such that a higher open loop UGF could be achieved. The modification consisted on implementing a second differentiator stage to increase the phase margin at the kHz regime. In addition to that, a low pass filter was implemented at the output of the

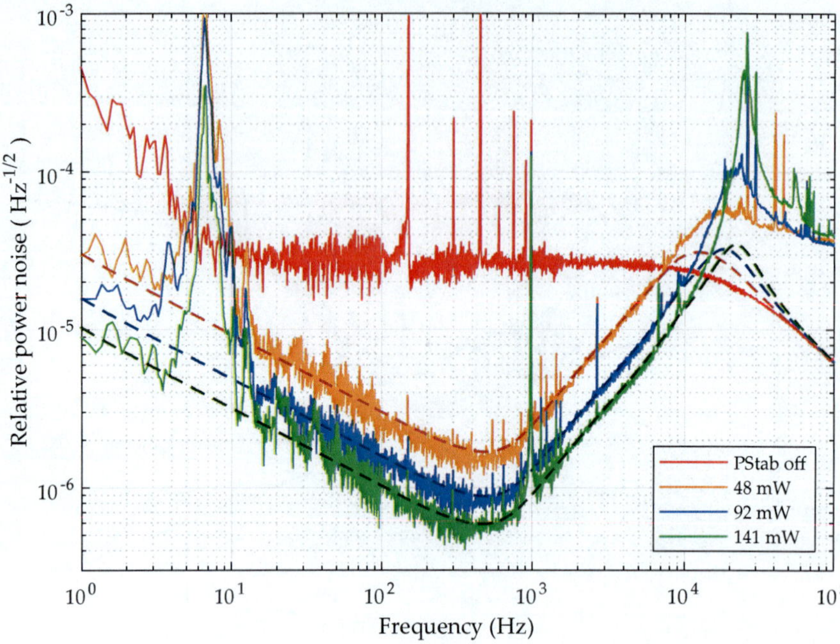

Fig. 5.24 ASD of the relative power noise measured by the out-of-loop sensor when the power stabilization control loop is turned off (red) and when the loop is on for different transfer beam powers. The dashed curves were calculated as an uncorrelated sum of the micro-oscillator's thermal noise and the expected free running noise reduction by the control loop. A broadband white noise was imprinted at the transfer beam in all measurements. The measurements were performed with a pressure of 10^{-4} mbar at the vacuum chamber, and with a UGF $\approx$ 9.4 kHz for a mean power of 48 mW, a UGF $\approx$ 15.5 kHz for a mean power of 92 mW, and a UGF $\approx$ 21 kHz for a mean power of 141 mW

controller to suppress the additional gain at frequencies around 200 kHz, mentioned in Sect. 5.4.1. With these changes, a maximum UGF of 21 kHz was achieved with a phase margin of 25°. Part of the results from this section are published in [117].

Figure 5.24 shows the relative power noise of the out-of-loop beam when the power stabilization control loop is turned off, and when the loop is turned on for different transfer beam powers. In all measurements, a broadband white noise was imprinted to the transfer beam. The noise was injected before the low pass filter implemented at the output of the controller, which explains the frequency dependence shape at high frequencies in the free running power noise curve. Apart from the 7 Hz peak, the power stability at frequencies below 500 Hz was thermal noise limited, and the RPN was linearly reduced by increasing the transfer beam power. This reduction in the relative power noise is expected since the absolute power modulation applied by the control loop to the transfer beam, in order to compensate the micro-oscillator's thermal noise motion and to suppress the sensing noise in the error signal, is independent on the transfer beam mean power. This would not be case if the sensing noise was limited by noise sources that depends on the transfer beam mean power, such as scattering for example. For frequencies above 500 Hz, the stability was limited by the control loop gain. All measurements were performed by keeping the same electronic gain in the controller, and by increasing the transfer beam power via an attenuation stage located after the AOM (see Fig. 4.4). In this manner, the transfer function from power modulation imprinted by the AOM to absolute power modulation after the attenuation stage increased proportionally with power. As a result, the magnitude of the open loop transfer function was also increased with power, which lead to a higher suppression of the free running power noise at frequencies above 500 Hz. An excess noise can be observed at frequencies between 10 Hz and 100 Hz, especially in the green curve, in which the transfer beam power was higher. The peaks present on this noise occurred at frequencies which are close to the harmonics of the 7 Hz peak, and were most likely caused by scattering from the transfer beam in the OOL PD, or by beam jitter at the photodiode. This explanation is supported by the fact that this noise was not present in the measured interferometer displacement noise, but it could be observed in the transfer beam free running power noise (without imprinting white noise) measured by the OOL PD (see Fig. 5.15). It is also interesting to note that the resonance behavior at the longitudinal resonance frequency in the free running noise measurement was particularly high for this oscillator, and that its corresponding 5 first harmonic peaks at 302, 453, 604, 755, and 906 Hz are also present. These peaks are not present in the measurements when the power stabilization loop is turned on because the micro-oscillator's longitudinal motion is reduced by the control loop gain.

The UGF of 21 kHz was close to the control loop instability point and could not be further increased. Hence, the stabilization for higher powers was performed by adjusting the electronic gain in the controller such that the open loop UGF was kept approximately at 15.5 kHz, as shown in Fig. 5.25. The curves now overlap in the region where the stability performance is limited by the control loop gain. At the maximum transfer beam power of 267 mW, a relative power noise of $3.7 \times 10^{-7}\ \mathrm{Hz}^{-1/2}$ was achieved at 250 Hz, corresponding to a free running noise reduction by a factor of 73. A significant difference on the stabilization measurement made with 267 mW can be observed: the 7 Hz peak contribution was two orders of magnitude smaller than with previous measurements. In addition to that, the excess noise at frequencies between 10 Hz and 100 Hz was not present at this measurement. The peak height was also reduced in the interferometer displacement readout. Since no modifications were made in the setup between the measurements, the cause for this difference is most likely related to environmental changes in the vibrations at the laboratory building, similar to the one observed in Fig. 5.9. It is worth noting that the alignment of the oscillator was not optimized in vacuum for the power of 267 mW, which can be seen by the coupling of the peaks corresponding to the pitch and yaw motion at the stabilized power noise measurement. This choice was made because at each step motion in the translation stage the micro-oscillator was ringing for a long time, which made the alignment a slow process, and also to avoid damage in the oscillator, since a large amount of scattered light was observed in the CCD cameras.

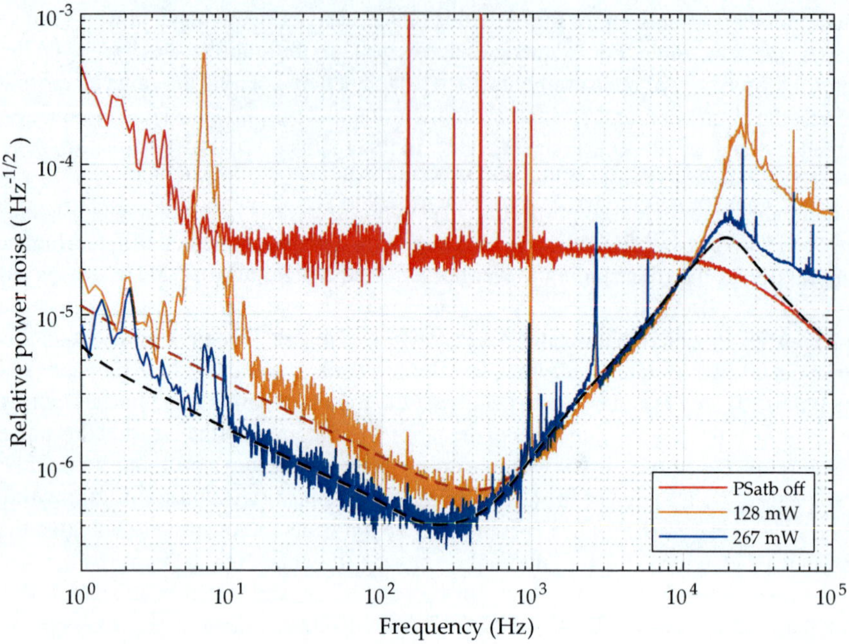

Fig. 5.25 ASD of the relative power noise measured by the out-of-loop sensor when the power stabilization control loop is turned off (red) and when the loop is on for different transfer beam powers. A broadband white noise was imprinted at the transfer beam in all measurements. The measurements were performed with a pressure of 10^{-4} mbar at the vacuum chamber and with a UGF of ≈ 15.5 kHz

For transfer beam powers higher than 200 mW, the free running absolute power fluctuations of the NPRO are large enough to be sensed by the interferometer for frequencies up to 1.3 kHz, such that no white noise needs to be applied in the transfer beam. This can be seen by the right plot of Fig. 5.26, which shows a comparison of the interferometer displacement noise (green curve) and the free running power noise of the transfer beam projected to displacement in the micro-oscillator (red curve). The left plot shows the performance of the power stabilization without white noise applied in the transfer beam. The stability is now solely limited by sensing noise, since the loop gain was enough to suppress the transfer beam free running power fluctuations. Thermal noise now limits the sensitivity until 1 kHz, and a power stability of 2.5×10^{-7} $\text{Hz}^{-1/2}$ was achieved at 730 Hz. At higher frequencies, the sensing noise is above the free running noise of the transfer beam and is imprinted on the out-of-loop beam following the frequency dependence of the radiation pressure transfer function. To reduce the imprinted noise by the control loop at high frequencies, a UGF of 10 kHz was set for this measurement.

The realignment procedure when increasing the power in the LC5 oscillator was performed with atmospheric pressure in the chamber, in order to reduce the risk of damage due to the large displacement of the oscillator induced by the translation stage

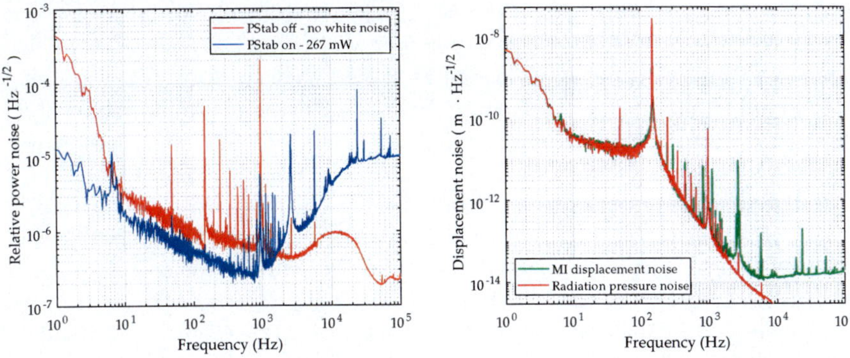

Fig. 5.26 Left: ASD of the relative power noise measured by the out-of-loop sensor when the power stabilization control loop is turned off (red) and when the loop is on, for a transfer beam power of 267 mW and without imprinting white noise. The open loop UGF was 10 kHz. Right: ASD of the interferometer displacement noise measured with a transfer beam power of 267 mW (green curve), and free running power noise of the transfer beam projected to micro-oscillator displacement (red curve). The measurements were performed with a pressure of 10^{-4} mbar at the vacuum chamber

step motion in vacuum. After each set of realignment steps, the beams were manually centered at the mirror pad in order to achieve a more precise alignment. With this procedure, the LC5 oscillator was so far not damaged, and could stand a power of at least 285 mW, which was higher than for the LA7 and LC6 oscillators. The power was not increased further in this experiment because the range in the pitch degree of freedom in the motorized mirror mount reached its limit, and it was not possible to continue with the realignment of the mirror pad. The static displacement of the oscillator at this power was calculated to be 11 μm, which represents a bending angle of 3.2° with respect to its relaxed position.[4] For this reason, the damage threshold of this oscillator was not measured.

5.5 Future Work

In this proof-of-principle experiment, an RPN_{ool} of 2.5×10^{-7} $\text{Hz}^{-1/2}$ was achieved at 730 Hz with the LC5 micro-oscillator. The stability performance at frequencies below 1 kHz was consistent with the expected thermal noise limit measured at room temperature. In summary, the results agreed with what was expected from the design of this experiment and from the sensing noise characterization. In future work however, a higher power stability in the out-of-loop beam is desired. Let us now discuss the possible steps to achieve this goal.

The straightforward path to achieve higher power stability with the current setup, at frequencies where the experiment was limited by structural thermal noise, is to further

[4] The angular range specified for this mount by the manufacturer is ±4°.

increase the transfer beam power. The power damage threshold for this generation of micro-oscillators has not yet been determined. However, measurements made with similar oscillators by our collaborators at the Louisiana State University set a lower limit for the damage threshold of 500 mW. This power would represent an improvement for the relative power noise by a factor of ≈ 2 with respect to the current result achieved with the LC5 oscillator. Another path would be to implement an oscillator with a mechanical susceptibility close to the LC6 oscillator, which would improve the stability level by another factor of ≈ 2. In this case, the alignment with high transfer beam power should be performed with atmospheric pressure at the vacuum chamber. The open question with both paths, that should be investigated in the future, is the damage threshold and also its breaking mechanism, which might depend on the oscillator's design. To achieve an $\mathrm{RPN}_{\mathrm{ool}}$ close to $10^{-8}\ \mathrm{Hz}^{-1/2}$ at 700 Hz, for example, a transfer beam power of 4 W is needed. Previous measurements of the damage threshold of a coating formed by 35 alternating layers of GaAs and $Al_{0.92}Ga_{0.08}As$, transferred to a few cm thick silica substrate, resulted in a lower damage threshold limit of $64\ \mathrm{MW}\cdot\mathrm{cm}^{-2}$ [118]. This would mean that powers on the order of kW could be used with the oscillators. However, an important factor to take into account is that the thick silica substrate supporting the coating helps with the transfer of the absorbed heat away from the mirror, whereas the free-standing mirror coating in the micro-oscillators is prone to more heating due to the weak thermal link between the mirror pad and the chip, which can vary depending on the design of the oscillator. Hence, an optimal design of micro-oscillators for a power stabilization experiment should be a compromise between a high mechanical susceptibility and a high power damage threshold, and should be investigated in future experiments.

At higher frequencies, the experiment was limited by a combination of the contributions from viscous thermal noise, sensing beam power noise, and electronic noise. By improving these noise sources, a relative power noise below $10^{-7}\ \mathrm{Hz}^{-1/2}$ can be achieved at the kHz regime with a power of 250 mW and the LC5 oscillator.

To achieve higher power stabilities, however, changes in the setup design might be required. An option that should be investigated, for example, is to use a self modulation scheme configuration. The implemented scheme based on a cross phase modulation was preferred for a proof-of-principle experiment over a self-modulation scheme[5] in order to have a clear distinction and independent characterization between the transfer and readout schemes. This was an important configuration to test the transfer concept discussed in this thesis. However, a self-modulation scheme could have advantages in a future experiment. One of the advantages is that the interferometer would operate with a beam power higher than 250 mW, which would increase the interferometer transfer function magnitude by at least one order of magnitude. As a consequence, the contribution from the MI control loop electronic noise to the sensing noise would be equally reduced. Another advantage is that technical power noise would not limit the interferometer sensitivity, neither by the mid-fringe

[5] In a self-modulation scheme, the transfer beam would be implemented in the interferometer and be responsible for both the phase transfer and the phase readout (see discussions in Sect. 2.2, and Fig. 2.10).

readout, nor by radiation pressure noise, since the power at the input of the interferometer will be stabilized by the radiation pressure control loop. However, with this configuration, a dark-fringe readout should be implemented in order to preserve the full transfer beam power to the out-of-loop beam. In addition to that, only one beam would need to be aligned to the micro-oscillator, which would simplify the alignment procedure. Another configuration that should be investigated is to perform the phase readout with a cavity. As discussed in this thesis, the in-loop signal for a cavity is in principle considerably increased with respect to an interferometer. In this case, a calculation to determine the stability requirements for the laser phase noise should also be performed.

Finally, to achieve lower power noise values, a cryostat that could reach a temperature close to 10 K should be implemented. An additional benefit of using cryogenic temperatures is that the structural quality factor of the micro-oscillator could potentially be increased by one order of magnitude [105]. Hence, a power noise reduction by a factor of 17 might be achieved by operating the micro-oscillator at 10 K.

Chapter 6
Summary

Abstract This chapter is a summary of the work presented in this thesis and of the future steps on experiments of power stabilization via radiation pressure.

This thesis investigated the pertinent question of whether active power stabilization schemes for high power lasers could benefit from the transfer of the power noise to a different observable that could be measured with less sensing noise and/or with less effort than by a direct detection of power noise. An analysis of different transfer scheme possibilities revealed that, with current technologies, most schemes are unsuitable to detect a relative power noise on the order of 10^{-9} $\mathrm{Hz}^{-1/2}$. One scheme, however, stood out for its potential in achieving a relative power noise stability below 10^{-9} $\mathrm{Hz}^{-1/2}$ and a generation of a strong bright squeezed beam: a phase transfer scheme via radiation pressure.

In the radiation pressure scheme, the power fluctuations of a strong laser beam are transferred to motion of a movable mirror. The mirror motion is determined by a Michelson interferometer employing a weak laser beam, which forms the in-loop sensor for the power stabilization control loop. Theoretical investigations showed that a phase transfer coefficient of 3.6×10^3 rad $\cdot$ W^{-1} can be achieved with a micro-oscillator mirror with a mass of 40 ng and a longitudinal resonance frequency 117 Hz. This coefficient is at least three orders of magnitude higher than what can be achieved with transfer schemes via the Kerr and cascaded Kerr effects employing nonlinear materials.

An in-depth noise analysis of the radiation pressure scheme was made. The calculations were performed considering that the interferometer containing the micro-oscillator is fundamentally limited by quantum and thermal noise. For a transfer beam with a power of 4 W, a relative power noise below 6×10^{-10} $\mathrm{Hz}^{-1/2}$ can be achieved for frequencies 10 Hz and 6 kHz. This value can be further reduced by increasing the transfer beam power and/or by increasing the mechanical susceptibility of the movable mirror. Furthermore, the calculations show a remarkable result:

M. Trad Nery, *Laser Power Stabilization via Radiation Pressure*, Springer Theses,
https://doi.org/10.1007/978-3-030-95868-8_6

a power stabilization below the shot noise limit can be achieved in the out-of-loop beam, with a squeezing factor of 11 dB, which can be increased with a higher laser power. This is a considerable advantage of the proposed scheme with respect to the traditional scheme assisted by squeezing, in which the squeezing factor is reduced with the out-of-loop beam power and would require 30 dB of amplitude squeezed vacuum in order to achieve the same power stability.

A proof-of-principle experiment with a design based on the theoretical investigations was realized with micro-oscillator mirrors with masses ranging from 25 to 190 ng, and longitudinal resonance frequencies from 151 to 219 Hz. The technical noise sources coupling at the interferometer readout were analyzed and reduced such that the interferometer was fundamentally limited by the thermal noise of the micro-oscillators at room temperature for frequencies up to a few kHz. A power stabilization was demonstrated for different transfer beam powers, and a relative power noise of $3.7 \times 10^{-7}\ \mathrm{Hz}^{-1/2}$ was achieved 250 Hz for a transfer beam power of 267 mW and for a micro-oscillator with mass of 190 ng and longitudinal resonance frequency 151 Hz. The out-of-loop power stability agreed well with the projection of the micro-oscillator's thermal noise to the out-of-loop beam at frequencies below a few kHz, and with the control loop's free running noise reduction at high frequencies. In addition to that, the expected improvement in the power stability with the transfer beam power was demonstrated. The maximum transfer beam power was set by a range limitation in the pitch degree of freedom of the micro-oscillator mount, which was not sufficient to compensate for the large longitudinal displacement induced by the mean laser power. Other experiments with similar mirrors have set a lower limit of 500 mW for their power damage threshold. Hence, an experimental investigation should be done in order to determine the damage threshold and also the breaking mechanism of these devices.

The successful implementation of the radiation pressure transfer scheme for laser power stabilization shown in this thesis paves a way for achieving a higher power stability in future experiments. An important step towards this goal is to design a movable mirror which is optimized for the purposes of a power stabilization of high laser power. The power tolerance of the mirror is a crucial aspect to be considered, since the power stability increases linearly with the transfer beam power. Another relevant parameter, but with a smaller impact, is the mechanical susceptibility of the mirror, which contributes to the power stability proportionally to the square root of its magnitude. Because the power tolerance can depend on the mirror design, a trade off between a high power damage threshold and high mechanical susceptibility is necessary. The investigations for a micro-oscillator consisting of a mirror pad and a cantilever should include, for example, an analysis of how the cantilever length and width, and the mirror pad radius can alter the damage threshold and the susceptibility. Other designs, such as a cat-flap type for example, or designs which the motion of the oscillator is constrained to the longitudinal direction, should also be investigated.

Finally, another improvement in a future setup is to reduce the thermal noise contribution to the interferometer readout. For the micro-oscillators used in this setup, structural quality factors of 1.7×10^4 and 6×10^3 were measured at room temperature. This value might be significantly increased with alternative materials and in a setup implementing cryogenics temperatures close to 10 K. Since thermal noise is a limiting factor in many high-precision optical experiments, the search for novel optical materials with higher quality factors is very active, and movable mirrors with high mechanical susceptibility might exhibit even higher quality factors in the future.

In summary, the results from the investigations performed in this thesis are a promising step towards a scheme that can achieve a relative power noise below $10^{-9}\ \mathrm{Hz}^{-1/2}$ at low frequencies and fulfill the requirements of future gravitational wave detectors. In addition to that, the scheme is a promising source for the generation of a bright squeezed beam, and can also be of interest for other high precision metrology experiments and optomechanical experiments.

References

1. Y. Wang et al., Shot-noise limited faraday rotation spectroscopy for detection of nitric oxide isotopes in breath, urine and blood. Sci. Rep. **5**1, 9096 (2015). ISSN: 2045-2322
2. Q.-X. Li, S.-H. Yan, E. Wang, X. Zhang, H. Zhang, High-precision and fast response laser power stabilization system for cold atom experiments. AIP Adv. **8**(9), 095221 (2018). https://doi.org/10.1063/1.5040238
3. N. Matsumoto, Classical Pendulum Feels Quantum Back-Action. Ph.D. thesis. University of Tokyo, 2016
4. G. Tino et al., Precision gravity tests with atom interferometry in space. Nucl. Phys. B Proc. Suppl. 243–244 (2013). Proceedings of the IV International Conference on Particle and Fundamental Physics in Space, pp. 203 -217. ISSN: 0920–5632. https://doi.org/10.1016/j.nuclphysbps.2013.09.023
5. P. Kwee, Laser Characterization and Stabilization for Precision Interferometry. Ph.D. thesis. Leibniz Universitaet Hannover (2010)
6. M.H. Poincaré, Sur la dynamique de l'électron, in *Rendiconti del Circolo Matematico di Palermo* (1884–1940), vol. 21.1 (1906), pp. 129–175. ISSN: 0009-725X
7. A. Einstein, Die Grundlage der allgemeinen Relativitätstheorie. Annalen der Physik **354**(7), 769–822 (1916). https://doi.org/10.1002/andp.19163540702
8. R.A. Hulse, J.H. Taylor, Discovery of a pulsar in a binary system. Astrophys. J. **195**, L51–L53 (1975). https://doi.org/10.1086/181708
9. G. Auger, E. Plagnol, An overview of gravitational waves. World Sci. (2017). https://doi.org/10.1142/10082
10. K.T. McDonald, What is the Stifiness of Spacetime? (2018). http://physics.princeton.edu/~mcdonald/examples/stifiness.pdf
11. P.R. Saulson, *Fundamentals of Interferometric Gravitational Wave Detectors*, 2nd. World Scientific (2017). https://doi.org/10.1142/10116
12. B.P. Abbott et al., Observation of gravitational waves from a binary black hole merger. Phys. Rev. Lett. **116**(6), 061102 (2016). https://doi.org/10.1103/PhysRevLett.116.061102
13. B.P. Abbott et al., GWTC-1: a gravitational-wave transient catalog of compact binary mergers observed by LIGO and Virgo during the first and second observing runs. Phys. Rev. X **9**(3), 031040 (2019). https://doi.org/10.1103/PhysRevX.9.031040
14. B.P. Abbott et al., Tests of general relativity with the binary black hole signals from the LIGO-Virgo catalog GWTC-1. Phys. Rev. D **100**(10), 104036 (2019). https://doi.org/10.1103/PhysRevD.100.104036
15. B.P. Abbott et al., Binary black hole population properties inferred from the first and second observing runs of advanced LIGO and Advanced Virgo. Astrophys. J. **882**(2), L24 (2019). https://doi.org/10.3847/2041-8213/ab3800

M. Trad Nery, *Laser Power Stabilization via Radiation Pressure*, Springer Theses,
https://doi.org/10.1007/978-3-030-95868-8

16. B.P. Abbott et al., GW170817: observation of gravitational waves from a binary neutron star inspiral. Phys. Rev. Lett. **119**(16), 161101 (2017). https://doi.org/10.1103/PhysRevLett.119.161101
17. B.P. Abbott et al., Multi-messenger observations of a binary neutron star merger. Astrophys. J. **848**(2), L12 (2017). https://doi.org/10.3847/2041-8213/aa91c9
18. B.P. Abbott et al., Gravitational waves and gamma-rays from a binary neutron star merger: GW170817 and GRB 170817A. Astrophys. J. **848**(2), L13 (2017). https://doi.org/10.3847/2041-8213/aa920c
19. B.P. Abbott et al., On the progenitor of binary neutron star merger GW170817. Astrophys. J. **850**(2), L40 (2017). https://doi.org/10.3847/2041-8213/aa93fc
20. B.P. Abbott et al., Properties of the binary neutron star merger GW170817. Phys. Rev. X **9**(1), 011001 (2019). https://doi.org/10.1103/PhysRevX.9.011001
21. M. Soares-Santos et al., First measurement of the hubble constant from a dark standard siren using the dark energy survey galaxies and the LIGO/Virgo binary-blackhole merger GW170814. Astrophys. J. **876**(1), L7 (2019). https://doi.org/10.3847/2041-8213/ab14f1
22. The LIGO Scientific Collaboration et al., A gravitational-wave measurement of the Hubble constant following the second observing run of advanced LIGO and Virgo. *arXiv e-prints*, arXiv:1908.06060 (2019)
23. B.P. Abbott et al., GW190425: observation of a compact binary coalescence with total mass ~ 3.4 M ⊙". Astrophys. J. **892**(1), L3 (2020). https://doi.org/10.3847/2041-8213/ab75f5
24. The LIGO Scientific Collaboration et al., GW190412: observation of a binary-black-hole coalescence with asymmetric masses. *arXiv e-prints*, arXiv:2004.08342 (2020)
25. M. Evans, J. Harms, S. Vitale, *Exploring the Sensitivity of Next Generation Gravitational Wave Detectors*. Technical report LIGO DCC P1600143-v18 (2016)
26. M. Punturo et al., The Einstein telescope: a third-generation gravitational wave observatory. Class. Quantum Gravity **27**(19), 194002 (2010). https://doi.org/10.1088/0264-9381/27/19/194002
27. D. Reitze et al., *Cosmic Explorer: The U.S. Contribution to Gravitational-Wave Astronomy beyond LIGO* (2019)
28. S. Vitale, M. Evans, Parameter estimation for binary black holes with networks of third-generation gravitational-wave detectors. Phys. Rev. D **95**(6), 064052 (2017). https://doi.org/10.1103/PhysRevD.95.064052
29. S.E. Gossan et al., Observing gravitational waves from core-collapse supernovae in the advanced detector era. Phys. Rev. D **93**(4), 042002 (2016). https://doi.org/10.1103/PhysRevD.93.042002
30. M. Bejger, Gravitational waves from rotating neutron stars: current limits and prospects. Phys. Particles Nuclei **46**(5), 839–842 (2015). ISSN: 1531- 8559
31. T. Regimbau et al., Digging deeper: observing primordial gravitational waves below the binary-black-hole-produced stochastic background. Phys. Rev. Lett. **118**(15), 151105 (2017). https://doi.org/10.1103/PhysRevLett.118.151105
32. G. Bertone et al., Gravitational wave probes of dark matter: challenges and opportunities. *arXiv e-prints*, arXiv:1907.10610 (2019)
33. L. Barack et al., Black holes, gravitational waves and fundamental physics: a roadmap. Class. Quantum Gravity **36**(14), 143001 (2019). ISSN: 1361-6382. https://doi.org/10.1088/1361-6382/ab0587
34. B. Willke, P. King, R. Savage, P. Fritschel, Pre-stabilized laser design requirements. LIGO DCC (2011)
35. T.T. Fricke et al., DC readout experiment in enhanced LIGO. Class. Quantum Gravity **29**(6), 065005 (2012). https://doi.org/10.1088/0264-9381/29/6/065005
36. M. Tröbs, P. Weßels, C. Fallnich, Power- and frequency-noise characteristics of an Yb-doped fiber amplifier and actuators for stabilization. Opt. Express **13**(6), 2224–2235 (2005). https://doi.org/10.1364/OPEX.13.002224
37. J.C. Alex Abramovici. *Feedback Control Systems* (Springer, New York, 2012)

38. F. Seifert, P. Kwee, M. Heurs, B. Willk, K. Danzmann, Laser power stabilization for second-generation gravitational wave detectors. Opt. Lett. **31**(13), 2000–2002 (2006). https://doi.org/10.1364/OL.31.002000
39. P. Kwee, B. Willke, K. Danzmann, Shot-noise-limited laser power stabilization with a high-power photodiode array. Opt. Lett. **34**(19), 2912–2914 (2009). https://doi.org/10.1364/OL.34.002912
40. P. Kwee, B. Willke, K. Danzmann, Laser power stabilization using optical AC coupling and its quantum and technical limits. Appl. Opt. **48**(28), 5423–5431 (2009). https://doi.org/10.1364/AO.48.005423
41. S. Kaufer, B. Willke, Optical AC coupling power stabilization at frequencies close to the gravitational wave detection band. Opt. Lett. **44**(8), 1916–1919 (2019). https://doi.org/10.1364/OL.44.001916
42. S. Kaufer, Optical AC coupling in the gravitational wave detection band. Ph.D. thesis (Leibniz Universitaet Hannover, 2018)
43. S. Kaufer, M. Kasprzack, V. Frolov, B. Willke, Demonstration of the optical AC coupling technique at the advanced LIGO gravitational wave detector. Class. Quantum Gravity **34**(14), 145001 (2017). https://doi.org/10.1088/1361-6382/aa7119
44. H. Vahlbruch, D. Wilken, M. Mehmet, B. Willke, Laser power stabilization beyond the shot noise limit using squeezed light. Phys. Rev. Lett. **121**(17), 173601 (2018). https://doi.org/10.1103/PhysRevLett.121.173601
45. V.B. Braginsky, Y.I. Vorontsov, K.S. Thorne, Quantum nondemolition measurements. Science **209**(4456), 547–557 (1980). ISSN: 0036-8075. https://doi.org/10.1126/science.209.4456.547
46. S.J. Cooper et al., A compact, large-range interferometer for precision measurement and inertial sensing. Class. Quantum Gravity **35**(9), 095007 (2018). ISSN:1361-6382. https://doi.org/10.1088/1361-6382/aab2e9
47. R.W. Boyd, *Nonlinear Optics*, 3rd. (Academic Press, Inc., Orlando, FL 2008). ISBN: 0123694701, 9780123694706
48. D.N. Nikogosyan, *Nonlinear Optical Crystals: A Complete Survey*, 1st edn (Springer-Verlag New York, 2005). https://doi.org/10.1007/b138685
49. C. Li, *Nonlinear Optics. Principles and Applications* (Springer, Singapore, 2017). https://doi.org/10.1007/978-981-10-1488-8
50. Z.D.C.H.B. Lene Vestergaard Hau, S.E. Harris, Light speed reduction to 17 meters per second in an ultracold atomic gas. Nature, 594–598 **397** (1999)
51. D.N. Christodoulides, I.C. Khoo, G.J. Salamo, G.I. Stegeman, E.W.V. Stryland, Nonlinear refraction and absorption: mechanisms and magnitudes. Adv. Opt. Photon **2**(1), 60–200 (2010). https://doi.org/10.1364/AOP.2.000060
52. A.A. Said et al., Determination of bound-electronic and free-carrier nonlinearities in ZnSe, GaAs, CdTe, and ZnTe. J. Opt. Soc. Am. B **9**(3), 405–414 (1992). https://doi.org/10.1364/JOSAB.9.000405
53. R. Quintero-Torres, M. Thakur, Measurement of the nonlinear refractive index of polydiacetylene using Michelson interferometry and z-scan. J. Appl. Phys. **85**, 401 (1999)
54. H. Cai, S. Liu, E. Lalanne, A.M. Johnson, Investigation of giant Kerr nonlinearity in quantum cascade lasers using mid-infrared femtosecond pulses. Appl. Phys. Lett. **106** (2015)
55. M.D. Levenson, R.M. Shelby, M. Reid, D.F. Walls, quantum nondemolition detection of optical quadrature amplitudes. Phys. Rev. Lett. **57**(20), 2473–2476 (1986). https://doi.org/10.1103/PhysRevLett.57.2473
56. Li, Z Xiong, G. Peng, P. Chu, Direct measurement of nonlinear refractive index with an all-fibre Sagnac interferometer. Opt. Commun. **136**(3), 223–226 (1997). ISSN: 0030-4018. https://doi.org/10.1016/S0030-4018(96)00689-X
57. H. Garcia, A.M. Johnson, F.A. Oguama, S. Trivedi, Pump-induced nonlinear refractive-index change in erbium- and ytterbium-doped fibers: theory and experiment. Opt. Lett. **30**(11), 1261–1263 (2005). https://doi.org/10.1364/OL.30.001261
58. G. Assanto, *Beam shaping and control with nonlinear optics, in Quadratic Cascading: Effects and Applications* (Springer, New York, 1998)

59. R. DeSalvo et al., Self-focusing and self-defocusing by cascaded second-order effects in KTP. Opt. Lett. **17**(1), 28–30 (1992). https://doi.org/10.1364/OL.17.000028
60. G.K. Samanta et al., High-power, continuous-wave, second-harmonic generation at 532 nm in periodically poled KTiOPO4. Opt. Lett. **33**(24), 2955–2957 (2008). https://doi.org/10.1364/OL.33.002955
61. M. Jazbinsek, L. Mutter, P. Gunter, Photonic applications with the organic nonlinear optical crystal DAST. IEEE J. Selected Topics Quantum Electron. **14**(5), 1298–1311 (2008). https://doi.org/10.1109/JSTQE.2008.921407
62. *Rainbow Photonics*. http://www.rainbowphotonics.com/. Accessed 2020-06
63. P. Günter, *Nonlinear Optical Effects and Materials* (Springer, Berlin, Heidelberg, 2000)
64. D.Y. Kim et al., Second-order cascading as the origin of large third-order effects in organic single-crystal-core fibers. Opt. Lett. **19**(12), 868–870 (1994). https://doi.org/10.1364/OL.19.000868
65. A.L. Belostotsky, A.S. Leonov, A.V. Meleshko, Nonlinear phase change in type II second-harmonic generation under exact phase-matched conditions. Opt. Lett. **19**(12), 856–858 (1994). https://doi.org/10.1364/OL.19.000856
66. G. Assanto, I. Torelli, Cascading effects in type II second-harmonic generation: applications to all-optical processing. Opt. Commun. **119**(1), 143–148 (1995). ISSN: 0030-4018. https://doi.org/10.1016/0030-4018(95)00312-V
67. G.P. Banfi et al., Frequency conversion through parametric interaction and cascaded processes in a NPP crystal, in *Conference Digest. 2000 Conference on Lasers and Electro-Optics Europe (Cat. No.00TH8505)* (2000). https://doi.org/10.1109/CLEOE.2000.910015, p. 1
68. Z. Wang et al., Cascaded second-order effects in N-(4-nitrophenyl)-l-prolinol, in a molecular single crystal. J. Opt. Soc. Am. B **14**(1), 76–86 (1997). https://doi.org/10.1364/JOSAB.14.000076
69. T. Corbitt et al., Squeezed-state source using radiation-pressure-induced rigidity. Phys. Rev. A **73**(2), 023801 (2006). https://doi.org/10.1103/PhysRevA.73.023801
70. R. Singh, G.D. Cole, J. Cripe, T. Corbitt, Stable optical trap from a single optical field utilizing birefringence. Phys. Rev. Lett. **117**(21), 213604 (2016). https://doi.org/10.1103/PhysRevLett.117.213604
71. D. Kleckner et al., Optomechanical trampoline resonators. Opt. Express **19**(20), 19708–19716 (2011). https://doi.org/10.1364/OE.19.019708
72. J.D. Thompson et al., Strong dispersive coupling of a high-finesse cavity to a micromechanical membrane. Nature **452**(7183), 72–75 (2008). ISSN: 1476-4687
73. J.D. Cripe, Broadband Measurement and Reduction of Quantum Radiation Pressure Noise in the Audio Band. Ph.D. thesis. Louisiana State University, Agricultural, and Mechanical College (2018)
74. E. Verhagen, S. Deléglise, S. Weis, A. Schliesser, T.J. Kippenberg, Quantum coherent coupling of a mechanical oscillator to an optical cavity mode. Nature **482**(7383), 63–67 (2012). ISSN: 1476-4687
75. K.A. Stankov, A mirror with an intensity-dependent reflection coefficient. Appl. Phys. B **45**(3), 191–195 (1988). ISSN: 1432-0649. https://doi.org/10.1007/BF00695290
76. Y. Yamamoto, N. Imoto, S. Machida, Amplitude squeezing in a semiconductor laser using quantum nondemolition measurement and negative feedback. Phys. Rev. A **33**(5), 3243–3261 (1986). https://doi.org/10.1103/PhysRevA.33.3243
77. A.E. Siegman, *Lasers* (University Science Books, Mill Valley, CA, 1986)
78. H. Kogelnik, T. Li, Laser beams and resonators. Appl. Opt. **5**, 1550–1567 (1966)
79. R.W.P. Drever et al., Laser phase and frequency stabilization using an optical resonator. Appl. Phys. B **31**(2), 97–105 (1983). ISSN: 1432-0649. https://doi.org/10.1007/BF00702605
80. E.D. Black, An introduction to Pound-Drever-Hall laser frequency stabilization. Am. J. Phys. **69**(1), 79–87 (2001). https://doi.org/10.1119/1.1286663
81. A. Thüring, R. Schnabel, Critical Kerr nonlinear optical cavity in the presence of internal loss and driving noise. Phys. Rev. A **84**(3), 033839 (2011). https://doi.org/10.1103/PhysRevA.84.033839

82. A. Khalaidovski et al., Strong reduction of laser power noise by means of a Kerr nonlinear cavity". Phys. Rev. A **80**(5), 053801 (2009). https://doi.org/10.1103/PhysRevA.80.053801
83. A. Thüring, Investigations of coupled and Kerr non-linear optical resonators. Ph.D. thesis. Gottfried Wilhelm Leibniz Universität (2009). https://doi.org/10.15488/7324
84. S. Saltiel, K. Koynov, I. Buchvarov, Self-induced transparency and self-induced darkening with a nonlinear frequency-doubling polarization interferometer. Appl. Phys. B **63**(4), 371–374 (1996). ISSN: 1432-0649
85. S. Louis, V. Couderc, F. Louradour, P. Faugeras, A. Barthélémy, Nonlinear polarization evolution in type I and type II second-harmonic-generation crystals applied to the mode locking of a pulsed Nd:YAG laser. J. Opt. A Pure Appl. Opt. **3**(2), 139–143 (2001). https://doi.org/10.1088/1464-4258/3/2/307
86. E. Hecht, *Optics* (Addison-Wesley, Fifth, 2015)
87. L. Lefort, A. Barthelemy, Intensity-dependent polarization rotation associated with type II phase-matched second-harmonic generation: application to self-induced transparency. Opt. Lett. **20**(17), 1749–1751 (1995). https://doi.org/10.1364/OL.20.001749
88. J. Milanovic, Generation of Nonclassical Polarization States of Intense Light using Photonic-Crystal Fibers. Doctoral thesis. Friedrich-Alexander-Universität Erlangen- Nürnberg (FAU) (2013)
89. J. Heersink, V. Josse, G. Leuchs, U.L. Andersen, Efficient polarization squeezing in optical fibers. Opt. Lett. **30**(10), 1192–1194 (2005). https://doi.org/10.1364/OL.30.001192
90. M. Lassen, M. Sabuncu, P. Buchhave, U.L. Andersen, Generation of polarization squeezing with periodically poled KTP at 1064 nm. Opt. Express **15**(8), 5077–5082 (2007). https://doi.org/10.1364/OE.15.005077
91. C. Kim, P. Kumar, Quadrature-squeezed light detection using a self-generated matched local oscillator. Phys. Rev. Lett. **73**(12), 1605–1608 (1994). https://doi.org/10.1103/PhysRevLett.73.1605
92. J. Milanovic, M. Lassen, U.L. Andersen, G. Leuchs, A novel method for polarization squeezing with Photonic Crystal Fibers. Opt. Express **18**(2), 1521–1527 (2010). https://doi.org/10.1364/OE.18.001521
93. C.M. Caves, B.L. Schumaker, New formalism for two-photon quantum optics. I. Quadrature phases and squeezed states. Phys. Rev. A **31**(5) (1985), 3068–3092. https://doi.org/10.1103/PhysRevA.31.3068
94. B.L. Schumaker, C.M. Caves, New formalism for two-photon quantum optics. II. Mathematical foundation and compact notation. Phys. Rev. A **31**(5), 3093–3111 (1985). https://doi.org/10.1103/PhysRevA.31.3093
95. S.L. Danilishin, F.Y. Khalili, Quantum measurement theory in gravitational- wave detectors. Living Rev. Relativity **15**(1), 5 (2012). ISSN: 1433-8351. https://doi.org/10.12942/lrr-2012-5
96. C. Gerry, P. Knight, *Introductory Quantum Optics* (Cambridge University Press, 2004). https://doi.org/10.1017/CBO9780511791239
97. M. Yanagisawa and J. J. Hope. Self-consistent input-output formulation of quantum feedback. Phys. Rev. A **82** (6 2010), p. 062109. https://doi.org/10.1103/PhysRevA.82.062109
98. M.T. Nery, S.L. Danilishin, J.R. Venneberg, B. Willke, Fundamental limits of laser power stabilization via a radiationpressure transfer scheme. Opt. Lett. **45**(14), 3969–3972 (2020). https://doi.org/10.1364/OL.394547
99. K. Jacobs, P. Tombesi, M.J. Collett, D.F. Walls, Quantum-nondemolition measurement of photon number using radiation pressure. Phys. Rev. A **49**(3), 1961–1966 (1994). https://doi.org/10.1103/PhysRevA.49.1961
100. Y. Yamamoto, H.A. Haus, Preparation, measurement and information capacity of optical quantum states. Rev. Mod. Phys. **58**(4), 1001–1020 (1986). https://doi.org/10.1103/RevModPhys.58.1001
101. H.B. Callen, R.F. Greene, On a theorem of irreversible thermodynamics. Phys. Rev. **86**(5), 702–710 (1952). https://doi.org/10.1103/PhysRev.86.702

102. H.B. Callen, T.A. Welton, Irreversibility and generalized noise. Phys. Rev. **83**, 34–40 (1951). https://doi.org/10.1103/PhysRev.83.34
103. P.R. Saulson, Thermal noise in mechanical experiments. Phys. Rev. D **42**(8), 243–2445 (1990). https://doi.org/10.1103/PhysRevD.42.2437
104. A. Cavalleri et al., Gas damping force noise on a macroscopic test body in an infinite gas reservoir. Phys. Lett. A **374**(34), 3365–3369 (2010). ISSN: 0375-9601. https://doi.org/10.1016/j.physleta.2010.06.041
105. G D. Cole, Cavity optomechanics with low-noise crystalline mirrors, in *Optical Trapping and Optical Micromanipulation IX. Proceedings of the SPIE* (vol. 8458, article id. 845807, 11 pp.) (2012). Society of Photo-Optical Instrumentation Engineers (SPIE) Conference Series (2012), p. 845807. https://doi.org/10.1117/12.931226.
106. N. Matsumoto et al., 5-mg suspended mirror driven by measurement-induced backaction. Phys. Rev. A **92**(3), 033825 (2015). https://doi.org/10.1103/PhysRevA.92.033825
107. K. Yamamoto et al., Quantum noise of a Michelson-Sagnac interferometer with a translucent mechanical oscillator. Phys. Rev. A **81**(3), 033849 (2010). https://doi.org/10.1103/PhysRevA.81.033849
108. M.B. Gray, D.E. Mcclelland, M. Barton, S. Kawamura, A simple high sensitivity interferometric position sensor for test mass control on an advanced LIGO interferometer. Opt. Quantum Electron. **31**(5), 571–582 (1999). ISSN: 1572-817X. https://doi.org/10.1023/A:1006991717589
109. J. Watchi, S. Cooper, B. Ding, C.M. Mow-Lowry, C. Collette, *A Review of Compact Interferometers* (2018)
110. H.M. Wiseman, G.J. Milburn, Squeezing via feedback. Phys. Rev. A **49**(2), 1350–1366 (1994). https://doi.org/10.1103/PhysRevA.49.1350
111. A.S. Villar, The conversion of phase to amplitude fluctuations of a light beam by an optical cavity. Am. J. Phys. **76**(10), 922–929 (2008). https://doi.org/10.1119/1.2937903
112. Newport. http://www.newport.com/. Accessed 2020-06
113. X. Ramus, Transimpedance considerations for high-speed ampliers. *Application Report SBOA122*. Texas Instruments (2009)
114. L.L.B. István, L. Vér (ed.) *Noise and Vibration Control Engineering: Principles and Applications*, 2nd (Wiley, 2005)
115. H. Goldstein. *Classical mechanics* (Addison-Wesley, 2002). ISBN: 978-0-201-65702-9
116. A. Abramovici, J. Chapsky, *Feedback Control Systems* (Springer, US, 2000). https://doi.org/10.1007/978-1-4615-4345-9
117. M.T. Nery et al., Laser power stabilization via radiation pressure. Opt. Lett. **46**(8), 1946–1949 (2021). https://doi.org/10.1364/OL.422614
118. P. Koch et al., Thickness uniformity measurements and damage threshold tests of large-area GaAs/AlGaAs crystalline coatings for precision interferometry. Opt. Express **27**(25), 36731–36740 (2019). https://doi.org/10.1364/OE.27.036731

Index

M. Trad Nery, *Laser Power Stabilization via Radiation Pressure*, Springer Theses,
https://doi.org/10.1007/978-3-030-95868-8